BY FRANCISC

THE MAGIC

OF THE

CRYSTALS

A TRUE ACCOUNT OF SACRED INITIATION IN THE BRAZILIAN JUNGLE

Originally published in 1994 by Circulo do Livra Ltda.
Sao Paulo, Brazil, under the title *O Mago Dos Cristals.*

© 1998, Francisco Boström

Library of Congress Card Number 98-066080
ISBN 1-886708-00-2
printed in the United States of America
10 9 8 7 6 5 4 3 2 1

Cover art by Mario Diniz Santos, Brazil

Cover Design by Morris Design, Monterey, California

Interior Design by Danielle Shillcock, Monterey, California

Edited by Ann West and Julie King, Carmel, California

Translated by Debbie Mounts, Sam Miquel de Allende, Mexico

MERRILL-WEST PUBLISHING
P.O. Box 1227
Carmel CA 93921
831 • 644-9096
e-mail: info@voyagertarot.com
web site: www.voyagertarot.com

CONTENTS

Disclaimer

The author of this book is relating his personal experiences and spiritual understanding to assist those in search of deeper meaning in their lives. Throughout, numerous rituals and ceremonies are described and could be potentially dangerous if not performed correctly. In the event you use any of the information in this book for yourself, which is your constitutional right, the author and the publisher assume no responsibilities for your actions.

Hymn to God

You who transcends all (by what other name could we call you?).

With what song could I praise you, if you stand above all?

By what word could I exalt you, if you can not be expressed in words?

Oh, You, who transcends all, who creates everything with your words.

How can our minds embrace you if you are unreachable?

Chapter I

El Anciano, The Wise Old One

I began to work with gemstones and crystals when I was 15 years old in the town of San Lorenzo in the State of Minas Gerais, Brazil. Most of my family lived in the region, and it was there that I met many people in this profession who had a great influence on me and my future. I became aware that working with precious stones held a real fascination for me, even though it was unstable work filled with danger and fraud.

I envied the life of those who worked with stones. The lack of a fixed working schedule and a permanent residence appealed to my youthful sense of adventure. These merchants traveled from place to place, living freely without country or borders. They didn't have to spend their lives working in tiny, cramped offices.

I was impressed with the gem workers' freedom from the sense of time so prevalent in our world. These people lived in a world apart, unbound by politics, sports or international problems. They surrounded themselves in stones, completely submerged in them. The only important thing in their lives was to discover more of these precious objects so they could feel their nearness and thus experience spiritual freedom.

Finally, thrilled by this strange world, I studied the profession and soon began buying and selling gemstones.

Entering this environment, I discovered a totally new world for myself—a world I thought had long ceased to exist—a world of small, dark rooms, similar to those of chemists in their laboratories, where quiet old men passed the hours with their eyepieces analyzing, selecting and preparing the precious stones. I observed whole groups of men struggling through the thickets in search of hidden wealth. These men were capable of selling their babies' milk in order to repay a debt or even killing over a trinket that someone may have robbed from them.

After only a short while living with such people, I was able to perceive the existence of a persistent underground occult atmosphere. Through a series of discrete signs, I became certain there was a profoundly powerful and militant magic beneath the facade of materialism.

From the very beginning, I was very aware of mysterious signals and esoteric symbols located behind doors or on the upper pages of notebooks; small altars erected in the back of houses where men worked the stones—altars where Baphomet and other ancient gods appeared rather than the Christ; and encoded words. I saw people who appeared skeptical but who, after a few drinks, began to talk about the hidden power of the stones and how, because of this power, they had been able to establish contact with gods and demons.

Later, in fact, I discovered entire secret societies in which many people, from the simplest hunter of stones to the millionaire, practiced a very ancient, potent kind of magic. It was precisely through one of these believers that I met my teacher, El Anciano, known as The Wise Old One.

On a trip I made into the interior of Minas to buy industrial minerals, I stayed in the house of a wealthy businessman named Jonas, who had been a friend of my grandfather. By day, Jonas worked very successfully in the precious stone business and by night, he dedicated himself to carrying out spiritual exercises on gravely ill persons.

During one of our evening conversations, I told him I was attracted to quartz crystals and that I intended to open a crystal store in Rio de Janeiro. Jonas seemed very interested in everything I told him and, over the course of several days, he brought up conversation, trying to understand my feelings for the stones and my intentions regarding them.

Our intimacy grew as I spoke of my studies in the ancient mysteries, and he ended up confessing that he was a witch. He told me about the spiritual ceremonies he carried out, revealing his association with the secret society of "The Brotherhood of Fire." This organization was headed by the teacher known as El Anciano.

The only thing I learned about El Anciano was that he was very wise, and that sometime in the past he had been extremely wealthy and had traveled throughout the world. Eventually, El Anciano had decided to abandon these pursuits and retire to an isolated place where he could dedicate himself solely to the understanding of magic.

Finally, Jonas invited me to seek membership in the Brotherhood. I accepted immediately and we set the date when someone would come for me to begin my "journey."

Jonas was so emotional as we said good-bye that I couldn't help but wonder if I were soon to be happily blessed by fortuitous circumstances or instead about to embark on a sojourn that would indeed end in my death. As it turned out, in some ways both assumptions were correct.

An old car, driven by an unpleasant person, came to pick me up on the appointed day. To begin with, we crossed the city of San Lorenzo and began climbing toward the high country in the distance. The car left the highway as night approached, whereupon we entered a deserted, seemingly unused dirt road. The road became more and more difficult to climb until we were able to move forward only very slowly, needing to stop often. Finally, it was as dark as the pitch of a thousand-year-old pine, and not one word had been spoken on the whole trip!

Our headlights barely lit up the way ahead and the noisy car slowed even further, trying to avoid obstacles as we wove along the narrow path. It was nearly two in the morning when we met a band of prospectors at a small campsite. The foreman, who had been waiting for me, offered something to eat and a straw mat to lie on. I fell into a deep sleep, despite the fact that eight or nine workers were playing a very noisy game of cards.

The following day, the foreman shook me awake quite early. The car was gone and in its place stood two horses. When I tried to tie my suitcase onto the back of one horse, the foreman laughed heartily and assured me that where we were going, I'd have no need of it.

We began our climb on a trail that played hide and seek, appearing and disappearing through the arid, tree-filled countryside. I was amazed to see large piles of crystals here and there and prospectors, looking for new deposits, were passing by with their burros. No doubt, this was a mining region.

We entered the forest and rode for more than two hours before arriving at an open spot on a slope, from which we could see a beautiful view of mountains. In the clearing was a small brick cabin near which were the ashes of a bonfire and nine stacks of almost transparent crystals that strongly reflected the sunlight. Further down, scattered about a garden, were more crystals shining.

When we reached the house, the foreman and I dismounted, and we stood looking toward the cabin, waiting. As the door opened, a slowly moving figure emerged. Upon seeing us, he came forward with energetic steps. The appearance of the old one was not what I had expected. His head was bent and his face was full of wrinkles, yet his eyes, so penetrating and riveting, exerted an enormous magnetism.

He came directly toward me and, firmly shaking my hand, spoke softly. "Hello! Welcome! I've been waiting for you."

When I turned around, the foreman had disappeared. El Anciano invited me into the house and offered a lunch of eggs and vegetables. We chatted about many unimportant things . . . the trip, my work, the prospectors in the area and so on. Finally, he said, "You must be exhausted. I'll show you to your room where you may rest."

The room was tiny with only a straw mat, a water pitcher and a bedside lamp. Not realizing how tired I was, I lay down and fell fast asleep. When I awoke, I didn't know what time it was because my watch had stopped. I was making my way in near darkness to the front door when I glimpsed El Anciano outside. He stared at me, his eyes brilliant in the twilight, and moved closer.

"We are going to talk," he announced, "and from now on your name will be Rama." At dawn, El Anciano was still telling me about my past, my present and my future. He revealed that he worked with the great power of the crystals during the "magic of the fire." This awesome force allowed him the ability to tap into universal knowledge, he said, and I had come to study with him.

"I am going to destroy all your worldly illusions, your ideas about yourself and every-thing else. Afterward, you may return again to your world to live in accordance with the new reality you have come to know." What I thought would last only a few hours or a few days continued for many months.

CHAPTER II

THE FALLEN GODS AND THE FIRE SPIRIT

The following day, I was able to observe peacefully the place I now found myself. There were crystals all about. El Anciano had half buried them by the thousands, forming a circle around the clearing, which could not be crossed by anyone unless he himself would give permission. Nine groups of crystals formed a star and a large triangle consisting of three furrows in the ground. At each point in the triangle was an enormous crystal geode. Even trees had crystals imbedded in their trunks, and there were more crystals strewn around all the plants.

The house was constructed at an angle to the four cardinal directions, and in each one of its corners was a large crystal point facing upwards. Just inside the door, nine crystals made the shape of a Tau (capital "T") on the floor, while still others seemed to form a pentagram. Likewise, in the nine points of the large circle outside, several crystal groupings created a series of sacred symbols.

All around the area I detected evidence of fire ashes, bonfire residue, metal torches, containers of petroleum, scorched kindling and plants, small rusted ovens and daggers with their blades buried in the dirt. It was in this atmosphere that I began my initiation into the Sacred Tradition. I was to be given the key to understanding the truth of the crystals, mankind, the universe and history in general.

According to El Anciano, knowledge about the magic powers of crystals is traditional wisdom that has existed from the beginning of time. Traces of these teachings can be found in the ancient cultures of Egypt, Tibet and Northern Europe. Nevertheless, the ancient wisdom has been preserved in its purest form only by the Chamanes of the Americas, who have passed it from generation to generation up to the present day.

"In all human traditions, written as well as oral," El Anciano informed me," we find reference to a great cosmic fall. The Sacred Tradition is based on the Myth of the Fallen Gods and the Great Sin." I followed the old man obediently as we walked toward the garden.

"There was a time when things were different," he went on to say. "The world was a happy place, with sublime energies that existed in harmony and unity. We were gods in this world. One day, a Great Sin was committed and everything changed. The gods lost their thrones and found the only way to perpetuate themselves in this new time and space was to transform into crystals." He told me that this magic had taken place through the Fire Spirit, and for this reason there is a profound relationship between crystals and fire.

The pure stones that today are called crystals are actually transformed gods. No wonder people have always been fascinated by these stones. To dominate them is synonymous with the domination of the gods, and it follows that he who is master of the gods can be the master of anything.

El Anciano continued to explain the way of the Great Fire Spirit. "It is he who provokes storms in both nature and mankind in order to halt self-aggrandizement. This spirit is everywhere—in the forests and the cities, in the sky and the land, in our surroundings and within. Mankind has given him many forms, but he is only one. He is the sacred serpent that bites in order to be reborn. He is the giver of light—the angel of life and of power—that burns in order to purify and set us free. Just as in giving birth, he brings us the light of a new life. He is the magic messenger that propels the evolution of mankind."

"How can we come to know him?" I asked.

"We already know the Great Fire Spirit, although at times we are not able to identify him. If you would pay more attention when reading history books, you would be aware of his presence, and if you would look around yourself and within, you would see him. He is everywhere."

"And how does he act?" I was incredulous.

"By confronting water, the rebellious force that works against evolution. There is a fight between fire—symbolized by the sun, dynamic and rejuvenating—and water—symbolized by ice, heavy and immovable. The clash of these two forces forms the language of the universe. This duality of evolution, however, is merely an illusion. There is, in reality, only one natural force that divides itself to permit the growth of all."

Later I learned more about this symbolic

language—fire representing the Spirit and water representing Matter or the Material. Then, El Anciano showed me the primary symbol of the Sacred Tradition—an X (a slanted cross) within a circle within a diamond.

"The diamond stands for the Sin and the Great Change. The point above represents the primordial unity in which everything existed. The left and right parallel points stand for the spiritual duality in the world while the point below represents the eventual return to unity once evolution is complete. The circle is the fire that energizes the X, or the water, so that it can unite with the oneness."

"And the crystals?" I asked.

"They are like the diamond, a powerful energy that accelerates evolution giving birth to matter! The crystallized gods wish to return to their state of being before the Great Sin, and that can only come to pass when the universe evolves to a great height. For this reason, the crystals are at the service of human will. They offer themselves to be used in a magical way so that Spirit may dominate the Material, helping mankind to advance."

"Why must this happen through the magic of the fire?"

"Because following the fire, the crystals recover their original power and often acquire a tremendous additional force. Whatever thoughts have been projected into them during the process are invariably realized. Inebriated by their own power, the crystallized gods thus become willing to serve anyone who brings about an awareness of this power."

Later, I asked El Anciano to tell me more about the initiation of the crystal fire. "The initiation takes place directly with the gods. The only thing that I do is transmit knowledge, waking up the consciousness of the internal fire and destroying the initiate's worldly illusions."

I was about to interrupt but he continued. "There are nine levels to the initiation. During certain periods, the initiation takes place within this world. In others, the person is isolated from the world. And in the final stages, the rituals occur in the 'other world.' The crystallized gods are our only teachers. They tell us what to do, when and how. If you open up to them, everything is very easy. As soon as you let go of your worldly illusions and are willing to destroy them, you'll be better able to understand it all."

I asked him if, once initiated, I'd have to quit smoking, drinking or other such things. "Of course not," he replied. "The great sorcerers drink, smoke, dance, make love, live life. The only difference is that they live more intensely than most other people. They are like the fire that conquer the water, activating it and giving it life. This is what the world is made of: blood, earth and struggles. And you, you are also all of these things. Don't hide from the world, don't be afraid. See it as it is, accept it, confront it. By living life to the fullest, you'll be helping the universe to evolve. It's not a matter of giving up the material life but of conquering it by immersing yourself in it through the Fire Spirit.

"There is a big difference between a sorcerer and a mystic. El mago," said El Anciano, "lives in the world confronting the reality presented to him while the mystic, forgetting that he, too, is human, tries to live above all this. The truth is that the fire conquers the water by impregnating it rather than by fleeing from it. This is the only way in which evolution can come about."

El Anciano could see my fatigue. He smiled and waved me toward the house where I gladly went. It seemed a blessing to have some time alone. Later that night, El Anciano stated that he was now going to destroy my worldly illusions. I would thus begin to understand that which could be known only by being absolutely conscious of a parallel and invisible reality.

He asked me to sit on one of two chairs at a small wooden table in the house. Silence and an impenetrable calm were all that reigned in this atmosphere. On top of the table was a crystal whose point was flattened so it could stand upright. There was also an empty glass holding a candle. Besides the tenuous light from the candle, the only other illumination came from a small lamp.

El Anciano told me to concentrate on the image of the glass reflected within the crystal. Afterwards, he remained still for a long time staring fixedly at the crystal without even once blinking. Time seemed to stand still, like a photograph. Suddenly, in a split second, El Anciano reached under the table, raised a hammer and smashed it into the crystal! Instantly, the crystal and the glass exploded, throwing out splinters which cut our hands and faces.

CHAPTER III

$$2 + 2 = 5$$

In spite of the newness of my experiences, I felt more rested the next day. El Anciano asked me to bring to the house a crystal geode that was nearby. The stone was very heavy and difficult to carry. Staring directly at me, without speaking a word, the old one indicated on which side of the floor I should deposit the crystal. Suddenly, the rock seemed to leap from my hands and a sharp point cut my finger. My blood stained the rock, and the contrasting red on top of the white stone was so fascinating I scarcely noticed my injury.

I turned toward El Anciano who said sincerely, "Congratulations. You have just been initiated by the crystals. They only manifest themselves once you've passed through a baptism by blood. Upon magically absorbing your blood, the crystal opens a pathway through your aura by which it can enter and begin to work with you. If it had not happened in this manner, you would have had to cut yourself and offer your blood to the stone. That is how, in the Sacred Tradition, one begins the magic work."

This first interaction with the crystals was so exciting that I began to ask questions about the initiation process. El Anciano seemed somewhat annoyed at having to talk again about this.

"The objective of the nine stages, Rama, is that the initiate is transformed into a living reincarnation of Dutomba, the first shaman of mankind. He will continue to live inside those who follow the Sacred Tradition to its final conclusion. Dutomba was a human incarnation of the Fire Spirit. Through the initiation process, you cease to exist and the Spirit enters to live inside of you. But you shouldn't worry so much about the initiation.

"Listen, just as I have with others, I've brought you here in order to pass on the Sacred Tradition. Later, anyone you do or don't initiate is your concern alone, and about this issue you must deal directly with the Gods."

I settled down more attentive than ever. El

Anciano and I moved outside, where we turned toward one of the piles of crystals gathered from the clearing. The moment had arrived for me to acquire my personal crystal.

"Pay attention. You will not be the one to choose the crystal. It will choose you."

I sat on a stone in front of the crystals with my hands on my knees, palms upwards. I closed my eyes and relaxed, clearing my mind of any thoughts. When I felt totally at ease, I was to open my eyes and take whichever crystal had an irresistible attraction, because that one would be, without a doubt, my personal crystal.

Upon opening my eyes, I immediately felt the magnetism of one of the crystals. It was mine! The stone transmitted to me such a sense of well being, pleasure and familiarity that I felt as if we were old friends.

El Anciano explained that a personal crystal is not used in magic ceremonies; rather, one treats it like the best friend one could ever have. It becomes a companion to protect us, keep us alert, inspire us to noble thoughts and bring good things into our lives.

"Have it always near you, in your hand or your pocket, and above all, never part with it. In time, your friendship will strengthen. You will speak with it, confessing your fears and asking for

advice. You will also be able to use it for thera-
peutic purposes, for healing or working with the
energy centers of the body (the Chakras).

"You must treat your crystal with love and
kindness. It will recognize your devotion and
respond likewise but with an even greater
intensity. If you ignore it or treat it with scorn, it
will treat you just as poorly, even to the point
where it may look for another owner who will
honor it as it should be honored.

"And don't let anyone else touch it. The crystal
so loves its owner that it is capable of self-
destruction rather than allow a negative force to
touch it."

Before long, I couldn't imagine when I hadn't
had my crystal. It truly had become a friend. I
was to find a name for it. According to the old
man, all crystals have their own name; but
because these names have an unintelligible
vibration for humans, the stones transmit to us
other similar names we can understand. He also
instructed me on how to find the best name.

> *In order to discover your crystal's name it is*
> *necessary to ask it directly. Supporting it with*
> *both hands, with eyes closed and great con-*
> *centration, ask the crystal if it would like to*
> *reveal its name. The first name, sound or*
> *expression that comes to your mind will be its*

sacred name. Knowing this name will make your magical relationship with the crystal even more intimate and powerful.

Next, El Anciano spoke about crystals in general. I found out that they are divided into three basic types: those with simple points; those with larger points, normally called generating crystals; and groups or clusters of crystals usually found in geodes.

Those with simple or small points are used more as personal crystals, although when grouped together they can also be used as auxiliary magic with the larger ones. It is these crystals that are appropriate to scatter on the ground to create a positive force around us. They were the same type El Anciano had spread on the floor and in corners of the house. He had dozens of them distributed around the furniture, his wardrobes and throughout drawers.

Crystals with large tips often have a base to support several large prongs and are used for the magic rituals of programming, evocation and celestial communicating. They should be kept wrapped in a black cloth, black being the color of power. Small crystals are especially for purification, protection and filling the environment with positive energy. Crystals with multiple points are like darts, sharpened for wounding or expelling negative energy. And the large ones, called geodes, are very special, to be used only in *ceremonies of high magic.*

Besides the three types of crystals, there is also a crystal called "tabular" that has a flat form. It is used in rituals as a magic sword, held in the right hand (the one that emits energy) to invoke celestial beings. This powerful transmitter of personal energy should be used when the bearer needs self-assurance.

Colored crystals are also used: amethyst, smoky, pink, green and yellow. These are stones with a specific purpose, and in the Sacred Tradition are considered to be lesser gods, although still very powerful. They should be respected but are not comparable to the white crystals.

My interest in crystals was growing moment by moment. "How can we clean the crystals if they become filled with negative energy?" I asked innocently.

"This is totally absurd! You can clean off dirt or dust—not negative energy! The Gods do not have negative energy," was his instant reply.

"Is there a way to give them more energy?"

"No, they don't need to be recharged. You can activate their powers by exciting them, like a woman excites a man, but you can't really recharge them. Crystals possess an intrinsic power totally independent of human action. Remember, they are gods."

El Anciano seemed a bit irritated with me as he looked into my eyes, "Rama, what you are trying to do is certainly in response to things you've heard about or listened to somewhere else. Don't pay any attention to them! They are a corruption of the Sacred Tradition, deceptions totally apart from the primitive knowledge. We usually call people who say such things one of the three I's: ingenuous, ignorant or idiots.

"This is also the problem with religious organizations. They try to structure that which cannot be structured and rationalize that which cannot be rationalized. The Sacred Tradition opens a much wider panorama—there are no simplistic formulas to follow.

"A relationship must be established directly with the gods. When you meet those who want to rationalize spirituality, try not to get close to them because they are probably just swindlers or persons who themselves have been deceived.

"The primitive magic is unique. It should not be intellectualized but simply carried out. The only thing you need is your intuition, your feelings and sensations. Reason doesn't count for anything. When you tell yourself that 2+2=4, you are imposing the limitations of absolute truth; but if you are capable of saying that 2+2=5, you will be in harmony with the Infinite. In fact, this is the only certainty in a world that is nothing more than an illusion. Whoever knows the invisible truths can taste a reality that is both dynamic and relative, a continuous process of transformation and evolution."

Then, smiling, I asked him about masculine and feminine crystals. El Anciano smiled back.

CHAPTER IV

THE FIRE RITUAL

Every Monday, at the same hour of the morning, the foreman who had brought me to the clearing came to speak with El Anciano. The man and his horse never once crossed the circle of crystals. They stood quietly, looking toward the house. If El Anciano needed something, he would direct his request to the man and he would answer. If the wise one had nothing to say, he would gesture to them from afar, and the horse and its rider would leave.

From my vantage point, I had become aware of the curious fact that not even animals dared cross the line of crystals. The place where El Anciano lived had become a scared temple respected by everyone.

I also saw how difficult it was to maintain the same schedule as El Anciano. He almost never ate and hardly slept. He spent every night engaging in nocturnal conversations with me until dawn, and then he would rest awhile. After a couple of hours, he would get up and work in his orchard, rearrange the crystals or go out to spend the day alone in the surrounding area.

We talked about everything—secret societies, history, politics, religion, the occult, philosophy. One of his favorite themes was secret societies and their place in the world today. He attached great importance to the evolution and transformation of these orders taking place in modern civilization.

One night when I was talking to him about my work and money, El Anciano brusquely interrupted. "Money, money! If people only knew it is nothing more than an illusion. Verily I tell you that he who has everything has nothing; and he who seems to have nothing has everything. I say this because at one time I had much money and I know what happens.

"I also once knew a stockbroker who was depressed about his financial situation. Because he had a wife and children, I felt sorry for him and taught him how to roll a bill around a crystal

. . . to visualize it multiplying. Soon he was very rich, with money raining on him from everywhere. However, his enthusiasm was so great that he began to abuse his knowledge. Eventually, his wife discovered his secret and began to copy his behavior. Intoxicated with their wealth, they soon separated—and his wife took all the money."

From El Anciano I learned about the "boomerang" effect of the crystals' magic. Whatever we affect through the power of the crystals will one day return to us many fold. If we use the stones to cure someone, we, too, will be cured some day. But, if we use the magic to destroy a person, we can expect to be destroyed. He said their force is so powerful, we must use them with much caution.

With this, El Anciano invited me to begin my first magic experience using the crystals. He proceeded to explain that everything in the physical world also exists in the form of an idea in the celestial world.

"In this manner, and because the crystals are really gods, all that is projected through one's thoughts becomes a divine command. The crystal transforms the projected thought into a powerful astral reality, which is then manifested quickly in the physical world."

I was to begin with a very simple program, the most elementary ritual that would begin at nine in the evening. I asked, "How will we know what time it is. We don't have any clocks here?"

"You simply need to look at the stars, Rama," answered El Anciano.

A little before nine o'clock, we sat down on the floor. El Anciano asked me to undress, saying that from this point on all the rituals would be conducted without clothing. In front of us, three candles formed a triangle. In the middle of the triangle was a steel container filled with kerosene and a large crystal. Below the base of the triangle was another container with a red rose in it. Everything was placed upon a black cloth that covered the floor.

"Just like in the Christian Mass, the most important part of the Fire Ritual is the sacrifice. Therefore, there is nothing more important, more just, than offering one form of life for another. It can be a flower or any other cherished object."

I had prepared myself for the ritual that day by meditating in the sun, bathing myself in the river, covering myself with the earth, then bathing myself again. I was ready.

I lit a bonfire and danced around it, jumping up and down and playing with the coals. Using words given to me by El Anciano, I blessed the objects to be used in the Sacred Ritual including the crystal within the black velvet. The moment the ritual was to begin, however, I didn't have the slightest idea how to start!

"That doesn't matter. It's only your first experience. Just think of something simple," said the wise old one.

"But what do you advise me to think about?" I stammered worriedly. "Should I think about the ascension, something spiritual, in some . . . ?"

"It is you who should decide. Think about something concrete, something you desire."

"Well, to tell the truth, what I really want is a woman!"

"Then, plan how to find one, now! The plan can take two forms. You can project what you want into the crystal through your thoughts, or you can give your desires a color or symbol."

At nine o'clock we lit the candles and took the crystal out from its cloth. I closed my eyes and tried to relax. When I felt totally calm, I opened my eyes and, contemplating the crystal illuminated by the candles, began to project into its center.

After a little while, and following instructions from El Anciano, I tossed some colored powder into a crevice of the crystal and some into the vase. The kerosene caught fire and burned the crystal while I kept tossing more powder onto it. After a few moments, my projected idea seemed to glow in the crystal and it lay thick in the air around as well. I tossed the rose into the fire where it crackled before being totally consumed by the flames.

As the fire died out, I waited for the crystal to cool. When I could safely touch it, I cleaned it with a handkerchief and wrapped it again. Getting up, I went over to the low bushes and buried the precious stone beneath their cover. Then, I returned and cleaned up the rest of the utensils.

"I thought the crystal would surely break apart in the fire," I commented later.

"Normally that doesn't happen. It depends on the ritual, what you are doing and the kind of energy with which you are working."

"And if it had cracked?"

"Well, it could have been either a good or bad sign, but in any case, you would have been guided by your intuition."

I told El Anciano that by the time the ritual had ended, the inside of the crystal showed cracks in the form of little triangular shapes.

"Yes," he said, "sometimes these shapes are incised in the crystal and other times they are shaded, like clouds. It means that the crystal is pleased and will finish the program."

"And what happens if it begins to split open?"

"It is absorbed by another . . . a god never dies."

The next ritual taught me what he meant. According to El Anciano, in this first stage I would be inducted into the magical world of precious stones, and from then on I would live a very different life, a life much more powerful than before. For this I would need to increase my defenses. So he demonstrated a special process to amplify the power of my personal crystal.

> *Put your crystal in the middle of a circle formed*
> *by nine crystals, their points facing outward*
> *toward you. Then, with a hammer, explode them*
> *one by one. The force of each broken crystal will*
> *be absorbed by the rest. When only yours is left,*
> *it will have the combined power of all the others.*

That night I asked El Anciano what would happen to those who had a crystal but did not know the teachings of the Sacred Tradition?

"Fortune will smile on them, they will receive the positive and curative powers of the crystal. But because they do not know all that the crystal possesses, they will not value it enough and the crystal will eventually end up in the hands of someone who understands its mysteries."

The next morning I awakened very worried.

My family had been expecting me to be gone only a few days and, although I'd hardly felt time passing when I was with El Anciano, I suddenly realized many days had gone by.

"I must call my family to tell them I'm all right."

The old man shrugged his shoulders. He signaled with his head that I could cross the circle without any problem and then went into his house.

It took a long time to reach the nearest town. From there I phoned home to say I was well and had been traveling with a friend to places totally new to me. I returned by the same path but with a great weight lifted from my shoulders.

I was thinking about the past few weeks and the incredible events that had happened. El Anciano was such a special person. His irresistible strength had attracted me like a magnet. The profoundness of his voice, his penetrating eyes, his teachings, his strange practices all had me completely wrapped up. The only thought, the only thing real and important in this environment was the unknown and how to discover it.

At a certain point in the road, I saw a beautiful young woman walking very slowly. She had a charming figure and beautiful black hair and eyes. Her abundant, pronounced breasts rose and fell beneath her blouse with every breath. She wore tight black shorts, and peeking out through her sandals were very lovely feet! She was like a brilliant apparition.

I walked nearer to her and she gave me a beautiful smile. With a strong miner's accent, I explained that I was a nephew of the foreman bringing him a message from his aunt. We walked back together, chatting the whole time about things in general but knowing exactly what we each desired. She smiled at all of my questions and also at all of my answers. After half an hour, we knew how many things we had in common. She wanted to travel—all the way to Rio de Janeiro, and everywhere in the world.

"You are very pretty!"

"So are you."

"You are as lovely as this whole region."

"Yes, this region is very beautiful and nearby there is a river. Do you want to see it?"

"Of course!"

We went to the river and sat on its banks. I kissed her lips and she fully responded, encouraging me. The grass served as our bed. We undressed and lay down on top of the soft bed of green. She was full of passion. We made love several times, accompanied by the gurgling of the river and the birds. Totally enraptured, we spent the rest of the afternoon together. As darkness came, she said a tender good-bye and left.

Running almost without breathing, I arrived home. I called out to El Anciano, "It's true, the program works!"

CHAPTER V

EL ANCIANO THE PERSON

El Anciano and I became accustomed to talking for hours and hours. He carefully selected a variety of topics, and I was spellbound listening to ideas that were totally different from any I had ever heard. One morning when we were in discussion near the clearing, El Anciano mentioned what he thought about organized religion.

"Yes, Rama, religions always pursue us. We know they began from the original black magic where it was first understood that gods communicate directly through humans. But the purity of the gods was lost in the humans, and only the Sacred Tradition was able to preserve this early knowledge. It also confirmed that men are free beings, thus posing a threat to organized religion. Modern religions transformed the magic into dogma, into rules and fixed teachings that became instruments of power in the hands of the priests. Because these men dominated the believers, they have been able to keep the true spirit from mankind."

"And how can this true understanding be freed for all to use?"

"By passing it from generation to generation as an oral tradition; thus, it retains its pure form without the pretense of rules. Religions may attempt to destroy this purity of spirit, but the true knowledge will last forever."

Without warning, El Anciano became quiet and very still. He looked intently toward the sky, as if he had noticed something strange. Suddenly, he began to convulse, seeming to be in pain. While putting a hand on his chest and one on his face, he took on an expression of fury, "Don't you hear it? They're here again. They've returned! I hear fearful growls through the branches of the trees. Damn them!"

I didn't hear a thing.

El Anciano rose abruptly and I followed him. In spite of his age, he began to run with unbelievable speed. Never had I seen him like this. After running for about twenty minutes, I heard the sounds of shots.

We entered the forest, El Anciano in the lead with the agility of a panther. All at once, he stopped and knelt down beside a hare that had just been killed by a bullet. It was a horrible sight. I looked at him and saw that he was crying. He picked up the hare into him arms, then dug a hole and buried the animal tenderly, putting near it a small crystal he was carrying. Afterwards, he recited a sad incantation in an unknown language.

"They don't kill for food," he repeated over and over, "They only do it for pleasure."

We continued walking through the forest until we came to a clearing ahead. I saw six armed men talking and laughing loudly. Ducking down so he wouldn't be seen, El Anciano moved forward as silently as a coyote before an attack. I came to his side and whispered, "Don't you see they are armed? Let's go back. Please, don't go on!"

But he signaled me to be quiet. My stomach went cold thinking that I was here, next to an old man ready to attack a group of armed men much stronger than we were. This was craziness! One of the men turned and saw us. All the others followed suit, and one of them muttered, "It's El Anciano!"

"Ungrateful ones! Accursed ones!" he kept shouting at them.

I touched his arm and implored, "Anciano, they're going to wipe us out; let's get out of here while we can!"

"Ungrateful ones! Accursed ones!" he yelled, waving his fists at them.

To my surprise, I perceived a look of fear, yes panic, in their expressions. The fattest one stammered, "But we are very far away. We thought here we could do it!"

El Anciano beat the ground with a wooden stick and moved toward the group. They dropped everything they had and started to run in all directions. He chased them with his stick raised, but not being able to catch them, he tossed the stick with such force it pierced one of the men in his back. The man staggered to the ground, then got up immediately and continued to run, not daring to look behind.

El Anciano stopped his pursuit yet kept on waving his fists and shouting expressions I was not able to understand. At last, we began our return. Now, however, El Anciano walked slowly, having resumed the same countenance as always. I will never forget with what interior strength this skeletal old man succeeded in conquering a group of armed men.

After this success, El Anciano passed several days almost without talking. He seemed to be very concentrated, going off alone all day. When he returned late at night, he would say very little and continue with his rituals.

Obeying his instructions, I spent all my time working with the crystals. I would concentrate while touching them, trying to overcome my uneasiness with them. After a few days, while searching for leaves to combust the fire Anciano built every night, I noticed a man hanging around near some of the crystals in the clearing. As I got near him, his face seemed vaguely familiar, and then I remembered I had met him some months earlier at a reception in Sao Paulo. He was one of the country's wealthiest men and a renowned gem expert. When he saw me, he also recognized me and asked, "Where is he?"

"I don't know. He left very early. I suppose he must be nearby."

"Well, he sent for me. I'm going to sit down and wait for him."

I stayed standing there not knowing how to react. "It's not necessary for you to remain here," he said, "You can continue with whatever you were doing."

Hours later El Anciano arrived. He went over to the visitor, shook his hand and took him into the center of the circle. When I started to enter the house, El Anciano signaled that I was to come into the circle, also. The visitor stopped and gave him a look to the contrary. But El Anciano spoke sharply, "He needs to be present. He is a part of this!"

El Anciano spread several maps on the ground that detailed the region north of Minas. Using instruments the visitor had brought to make precise calculations, the old one revealed the exact location of an enormous secret deposit of crystals. The two men then said good-bye laconically and the visitor left.

I found myself very confused. A few days before, El Anciano had risked his life to stop the killing of some animals, and now he was pointing out the location of a mine from which men would extract millions of crystals to be destroyed in high-tech industries. He explained this seeming paradox to me.

"Rama, this is my great vengeance and also my great goodness. By helping men to discover the crystals buried underground, I collaborate with the universe. Some of these crystals will be used for personal growth, while others will be destroyed for money. And this duality speeds up the events in respect to the great changes that must occur.

If you observe the recent history of mankind, you'll see that the big events taking place in the world have always coincided with an increase in the extraction of crystals. This is evolution, my son, and we cannot flee from it. The world, just as we know it, is on the verge of ending, and the goal of the crystal gods is to accelerate this transformation."

"And how do the crystals help?"

"By accelerating the arrival of the earthly reign of the Great Fire Spirit. This reign will be much better than what the world has experienced up to this point."

"But with so much extraction, isn't there the possibility of depleting the crystals?"

"Those that are discovered represent a mere fraction of the quantity of crystals still buried in the earth. Besides, keep in mind that the remaining crystals will absorb those that have been destroyed, making themselves even more powerful."

We stood together without speaking until I could no longer contain my curiosity. I asked him, "Anciano, how did you know where the mine was?"

"The crystals told me."

The enigmatic part of El Anciano intrigued me. Although he was friendly and attentive, he was just as capable of suffering sudden attacks of rage. Although he was used to talking a lot, from time to time a profound and impenetrable silence overtook him. He was generous and attentive to my well being, but with regard to the discipline of the exercises, he was immovable. All of this duality brought to mind the words of a good friend who used to say that magic comes from the equilibrium between strictness and mercy.

Another great contradiction of the wise old one was his relationship to nature. He always referred to nature poetically, was capable of sitting totally enraptured watching a sunset and demonstrated a true adoration for animals and plants. Yet, when it rained, he was outraged. During a downpour, he would go outside and angrily shake his fists at the sky. Afterwards he would come back completely soaked, cover himself with a shawl and remain silent, staring at the fire.

One time he mused, "You must think I'm crazy, huh, Rama?"

Not waiting for an answer, he replied. "When the visible realities mix in your mind with the invisible, all points of view merge to such an extent that people think you've gone crazy. But, really, it's the crazy ones who live in the world."

The Fire Ritual El Anciano practiced every night seemed to be the most important thing in his life. The ceremony, always beautiful, long and complicated, would begin at eleven o'clock and continue until dawn.

During the first weeks of my stay, El Anciano compelled me to remain inside the large circle in the clearing, intending to protect me from the powerful energies of ritual. With the passing of time, however, I was permitted to observe his ceremonies from within a smaller magic circle, situated near to where the ritual took place.

El Anciano would first boil water in a pot filled with crystals. Then he chanted magic names while pouring the water around the circle where the ritual was to be performed. Within this circle, he made ready a huge fire using dry branches and lots of charcoal, into which he placed enormous crystal geodes. Afterwards, with the point of a crystal, he would draw a large circle and write the fifty-four sacred names associated with the Fire Spirit.

He always threw a red shawl on his back, wore a gold medallion embossed with the symbol of the Sacred Tradition and went barefoot, even on the coldest of nights. He also wielded a large crystal point as a rod and carried a steel dagger with a black handle studded with a brilliant crystal. He would begin the Fire Ritual by circling the dry branches, intoning old chants that spoke of the Great Fall, of the gods transformed into stones and of the hope of returning once again to the primordial unity.

With his rod, El Anciano would next trace signs in the air in the direction of the fire. Wielding the rod, he would point toward the sky evoking the Spirit of the Fire, the gold serpent, the arrogant lion and the great red sun. Then he'd drench the bonfire with kerosene and, with a match, light it.

As soon as the fire began to illuminate the night, El Anciano started the magic procedures through which he was able to communicate with the Fire Spirit. He now could make contact directly with any astral entity he wished.

According to El Anciano, these rituals are like daily masses offered to the great Fire Spirit for the evolution of mankind and for the strengthening of his own magic power. He claimed they influence the astral realms, exerting a great influence over the course of events in the world of man, especially now that the evolution of the universe is accelerating. In compensation, the Fire Spirit transmits to him knowledge, power and wisdom.

At the end of each Fire Ritual, El Anciano

gathered up the ashes and flung them into a stream. "Why do you do this?" I asked him once.

"The streams flow into rivers and the rivers into the oceans which, in their turn, distribute the effects of the Ritual to all the continents."

Chapter VI

The Ritual of Ecstasy

"In order to have power over the crystals, it's not enough to simply know how to use them," affirmed El Anciano. "Above all, Rama, you must establish an emotional and intimate relationship with them; the more profound this connection, the greater will be your power."

With the idea that I could develop this kind of relationship with my crystals, El Anciano decided to initiate me into the Ritual of Ecstasy.

"The ritual, aside from being a euphoric contemplation of the crystal, will unlock the latent magic prowess that resides in all men— powers such as telepathy, clairvoyance or pre- monitions. The ceremony should be divided into three parts, which help purify the aura and allow one to become united with the crystal."

El Anciano made me sit down while he rotated the point of a crystal around my head. Continuing, he showed me a method to clear and open the Chakras. "This exercise must be done before any kind of work with the crystals," he warned.

> *You hold a crystal in each hand, in the opposite direction of the wrist. Then, with your arms wide open so that your body makes the form of a cross, make rotations of 180 degrees, without breaking the straight lines of your outstretched arms. These movements should be repeated rapidly for several minutes.*

Then, El Anciano took out a beautiful crystal point from a piece of black velvet. It was about 15 centimeters long and contained many crystal-lizations.

I took the stone, walked up to the place where the stream began and sat down on a rock. Just as El Anciano had recommended, I closed my eyes and tried to clear my mind of any thoughts. After my mind had relaxed and I was focused, I opened my eyes to admire the crystal.

I realized I had never before achieved a real appreciation of any crystal. All of its angles shone brilliantly in the sun. I was captivated by its beauty, although my eyes could hardly assimilate the fantastic, infinitely multifaceted image it presented.

Within the crystal, I discovered a new universe; I became like an astronaut exploring other worlds, strange and mysterious. I could visualize images reminiscent of buildings with a thousand floors, extraordinary walls, waterfalls, clouds, valleys and precipices. The crystallizations took the shape of hands, animals, instruments, symbols and sacred signs. I could also see balloons, rainbows and galaxies of stars; images of solitude, immensity and grandness throughout horizons; and panoramas of indescribable beauty. There were countless doors leading to other worlds, and I was filled with awe.

As I moved the crystal, new images would appear—shadows, cities of clouds or pyramids floating in the air. Its capacity for variety was endless, far beyond any concept of time and space I had known. Gazing at the crystal, the whole universe seemed to be passing before my eyes.

Without warning, I felt I would topple from a sensation of vertigo. It seemed I was falling into an endless well, then into infinity. I was light, sensing nothingness—no ego, no boundaries. The idea of who and where I was disappeared.

Everything was only the immensity of this depth, nothing else. It was then that I understood why El Anciano would always say, "In everything is nothing, nothing is everything, and this is God."

Suddenly, a strange sensation reminded me that I still existed, that I had a body. But, this feeling was fleeting, and once again I began to sink into nothingness, submerging myself even deeper, almost totally. I sensed extreme heat over my head and upon my face. It was this realization that brought me back to consciousness. Besides feeling hot, I began to hear a noise, something similar to a voice, a call from very far away. It was beckoning someone, but who?

Then the light of the intense heat shocked me to my senses. I felt my own existence, recognized myself and realized that I was the one being called. When I opened my eyes, near my forehead was a candle that was almost burning my face. El Anciano was holding it. All around me it was dark; night had fallen.

"Calm down, Rama," El Anciano said. "The ritual has not yet ended."

We returned to the house, and in spite of having a slightly singed face, I felt very light. El Anciano began to speak. "You went too fast in the first part of the ritual. But since this deals with an initiation of fire, you'll still have to finish the rest of the steps tonight. What are you feeling?"

"Curiosity and fear."

"Wonderful! The curiosity will cause you to continue and the fear will protect you."

It was nearly as cold inside the house as it had been outside. The fire was out in the chimney and there was only one candle burning on top of the table. El Anciano removed another crystal from black velvet. This one, however, did not have crystallizations in its center. It was completely transparent.

El Anciano started a cozy fire and I began the second stage of the ritual, meditating on the candle's flame through the flawless crystal. The images I began to visualize were totally different now, yet they were equally as compelling.

My intimacy with the crystal deepened by the minute, and it wasn't long before I was hypnotized by the candle. Still, I was overcome by fear. How was I to be in the meditation and not completely lose myself? Even though he had said not to, I couldn't help blinking my eyes; the temptation was too strong. Then El Anciano emitted a series of noises and unconnected words that propelled me back to the room.

This was a real test of self-domination, to permit myself to meditate but to not become dominated by the act. I tried again. After some time, I became aware that I was achieving a perfect balance between a state of consciousness and non-consciousness. El Anciano also noticed this was happening and extinguished the candle. Thus, the second stage was at a close and we began to initiate the third stage of the ritual.

"This afternoon you learned to immerse your mind in the crystal, Rama, and now it is the Fire's turn. This will be your first contact with the Great Fire Spirit."

We sat down at the hearth of a roaring blaze. Looking directly into the flames, El Anciano began to talk. He seemed fascinated by the warmth of the fire that brightened his face.

"Look how beautiful the fire is! It was dark and the fire brought us light. It was cold and it provided heat. There was silence and it pleased us with its mysterious voice.

"Wisdom resides in the fire. Its flames show us how life is continuously improving itself. Its crackles emulate frequencies of the evolving universe. The fire that delivers, destroys and brings happiness to life lies dormant within each of us. If we succeed in tapping this force, we can transform the world.

"The fire is the divine emissary of the Supreme and Great Fire Spirit. Through it, the crystals exercise their magic powers with such ability that they manifest human desire into reality."

For hours, El Anciano talked about the Fire Spirit. Several times he rekindled the embers until they burst again into dancing fingers of light. Intoxicated by his words, I had to focus intently on the flames in order to stay present. It was well past dawn when he began to intone a haunting chant. Opening a little velvet bag, he threw a yellow powder into the fire and stepped away from the furnace.

The flames crackled fiercely. I sensed a presence in the house; something powerful floated in the air. A strong pain gripped my chest and I had trouble breathing. The power was invisible, but its presence was so tangible I could almost touch it.

CHAPTER VII

THE QUINTESSENCE

"Anciano, I believe we're going too quickly."

"You're right, but it's necessary."

I felt somewhat bewildered and even frightened by the teachings and experiences; I was losing my worldliness. The presence of the Fire Spirit was very awesome; however, the force was so strong I was overwhelmed by its power. I was not prepared or ready to live with it.

"This is only a question of time, Rama. When your worldly ways are totally destroyed, you'll feel so comfortable with the Fire Spirit that you'll invite its presence."

"This may be the case, but I'd prefer to go more slowly," I insisted.

"No! It can't be that way! The more forceful your journey into this world, the stronger you'll be as you leave it. You can achieve this only by breaking all your mental structures, the Western mind with which you were born and bred. You must renounce the water and submerge yourself totally in the fire in order to become incandescent, ready to serve the evolution of the universe."

Convinced, I surrendered. "Alright, Anciano. Help me, then, to go through the necessary stages."

"Rama, the moment has arrived for you to pass through the Quintessence. This is akin to the true reality of the world, to the natural state of things in the astral worlds rather than how things appear in their physical state."

"And how will I get to the Quintessence?"

"By introducing into the fire a specially prepared magic crystal." This is all he would tell me.

For three days, I prepared rigorously. I fasted, used several techniques to purify my aura and meditated on the fire by looking through a crystal. At night, I slept in a specially designed circle of crystals. My final preparation consisted of a small Fire Ritual in which I programmed a vision of the Quintessence.

On the morning of the fourth day, El Anciano took from the hearth of his oven a sturdy piece of wood and a small crystal.

"This crystal will allow you to see things very close to your Quintessence. If you remain alert, you'll be able to direct the sensations yourself instead of having them manipulate you. When the image of the Quintessence surpasses your ability to absorb it, you must avert your mind or the force will return so strongly it could drag you down, even to the point of death. But don't be afraid. Dominate it! Don't lose consciousness, and learn to master your thoughts."

He then had me take the crystal with my left hand and instructed me to walk wherever I wanted. "This is all that I have to do?" I asked, frightened by the simplicity of the instructions.

"That's it. Now begin!"

I could not understand how such an easy thing could give me access to something so important as the image of the Quintessence. But, seriously, I began to walk. I traveled around the house several times and then headed in the direction of the stream. Climbing to its head-waters, I was relieved to find everything quite normal. But, suddenly, I looked at the water and felt my world whirl. I righted myself and continued walking until I glimpsed a strange light that seemed to come from a spider in its web.

"It can't be a reflection from the sun," I thought. "It's too strong." I stood staring at the spider and its web. I felt the light getting brighter until the spider grew so big there remained only a small dot of light, but one so strong it hurt my eyes. I looked away and moved ahead, but everything around me now appeared to be different. The colors especially were more intense.

I tightly closed my eyes, and when I opened them again all was normal. With relief I went ahead, stopping only to admire an incredibly white flower that had caught my attention; but soon its brilliance scattered so forcefully that everything around it was erased. A bit startled, I blinked in earnest until the scene had returned to its place.

I finally stopped to rest while observing a tree. In an instant it disappeared, and in its place were hundreds of small circles superimposed upon each other in the shape of a cone. There were millions of dots of light of every color moving frenetically, one on top of the other, joining and separating continuously.

This extraordinary sight was accompanied by a sickening headache, but I was able to avert my gaze. All around me now, everything had been transfigured and exuded a tremendous energy.

I was hypnotized by darting lights wherever I looked. This new vision distressed me so much I tried to escape by looking toward the sky. But I couldn't get away from it. There were millions of absolutely brilliant blue dots moving rapidly. I was on the verge of fainting, my heart beating so strongly that I had to support myself on the trunk of a tree in order not to fall down.

I didn't know where to look. Totally lost among all these impulses, I walked toward the road, now transformed into a kind of elongated tunnel. The only thing I wished for was to get away from there. I arrived at the stream, but this had changed into a raging torrent, with gigantic waves crashing on top of each other as if desperately trying to reach some place else.

At the bottom of the river, besides fish I could see millions of spiny, agitated creatures. Now, the pebbles had become enormous rock formations in which coursed an infinite number of atoms come to life. Glancing toward the depths, I saw a sea serpent, completely still, staring back.

I perceived the creature's magic essence and felt a power emanate to the most profound core of my being. My essence seemed to teeter on the edge of a cliff, and I felt myself fall and fall. In desperation, I remembered the words of El

Anciano: "Switch your thoughts, switch your thoughts. Don't let them drag you down!!"

Then the abyss disappeared, and when I opened my eyes I saw the stream, serene and friendly as always. Trying to calm myself, I focused on myself. Sitting on a rock with my head in my hands, I didn't want to see anything else. I stayed like this awhile, not daring to look around.

But something caught my attention; it was my own body. Blood raced through my veins from my head to my feet. My heart pounded mightily like a bomb exploding repeatedly, agitating my whole being. The impression that all my organs were alive matched the sensation that my brain too had life. I must return home and beg El Anciano to stop all these hallucinations.

I opened my eyes, got up and ran all the way to the house, trying to look only at the ground beneath my feet. But this also disappeared, and in its place was a gigantic black canyon. I slipped into it, engulfed by many channels filled with water and boulders. I continued my descent to the center of the earth until I hit the ground.

There wasn't anything to do. Wherever I focused, I met horrible nightmares. Suddenly, I remembered the crystal in my hand; I would throw it far away and be done with all this. But instead of finding the crystal, out came the strongest vision of all; I trembled under a volcano insanely spewing fire in all directions and lava leaping all around, forming an incandescent lake. The smoke and sulfur were choking me. I had almost lost consciousness when El Anciano appeared, opened my hand, took out the crystal and helped me up.

Everything had now passed by and the colors returned to their normal state; from the green of the leaves to the blue of the sky, all was restored. Terrorized, bruised and exhausted, I couldn't help feeling angry at him while he was leading me home.

"Why did you let these things happen? I'm smashed to bits!"

"It was necessary for you to pass through the tunnel, Rama. You had a vision of nature without its masks. Although that's not quite the same as a vision of the Quintessence, the effects have been the same. You will understand everything when we complete all of this."

"And when are we going to finish?" I asked, on the verge of collapse. "You mean there is more?"

"Of course!" he answered. "Next, you'll learn what is inside of you. It's necessary that those things you were trying to forget or ignore all these years come to the surface. You'll have to let them flourish and confront them. Only in this way will you acquire sufficient strength to submit to understanding."

The last stage was again deceptively simple: A crystal prepared by El Anciano would allow knowledge, usually hidden by the mind, to come forth. I needed only to place the stone next to me while I slept. To my horror, however, this stage was even more challenging than the one before.

Vertigo would soon penetrate my emotions and intense pain latch on to my feelings. During the night, I had dreams in which I could re-member many forgotten childhood traumas, senseless insecurities, sequestered fears; all the frustrated illusions that had profoundly marked me accepted the invitation to come forward.

I met people I had forgotten. I was able to speak with ancestors who helped me see things I had previously denied. I recognized the weakness from which I had fled and my hopes of achieving things I really didn't desire. I saw myself as very young and cried for that small, innocent creature who was exposed to an inexorable world. Finally, I witnessed myself as I was, stripped of fantasy, illusion or disguise.

When I awakened, I felt totally transformed. I was much more skeptical about life, but also more confident now that I knew myself better.

"Congratulations, Anciano! You told me you would destroy my worldly illusions and you have finally done it. But why do I feel so empty? You ought to replace them with something else."

"It will be you, yourself, that will accomplish that. I've only given you sufficient equilibrium to do it alone."

I began to understand. The hate I had felt toward El Anciano had vanished and in my heart was a sense of peace—and immense gratitude.

CHAPTER VIII

TECHNIQUES AND CUSTOMS

After having destroyed my worldly illusions, El Anciano spent much time teaching me all the details of how to do magic with the crystals. I was, at last, learning the techniques and customs of the Great Tradition.

The first step, before starting any ritual, is to pass the crystals nine times through the fire in order to activate the power of the gods. Secondly, one must fast for at least six hours before working with the crystals, and it is important to disrobe in order to be alert and totally present. The third requirement, which is the most important, is to trust one's intuition. Although the Sacred Tradition recommends certain formulas when performing rituals, it is always up to the initiate to decide how many hours to spend or which words to use in the chants.

El Anciano also taught me how to cleanse my aura of negative energy in order to create a neutral, clear space for the magic to work.

> *Cleansing is done by rotating a crystal clockwise over your head. The proper way to hold the crystal is with your index finger and thumb, grasping the middle with the point facing toward the ground. By doing this, all negative energy contained in the aura is expelled. It is a very simple exercise that lasts about five or ten minutes. It will revive you after a difficult day or when you have been surrounded by crowds or in touch with negative influences.*

Another simple technique for cleansing the aura and the Chakras is to hold a crystal with its point away from the wrist and move it repeatedly up and down while encircling the upper part of the body. This exercise will create a strong electromagnetic field of protection.

As I listened to my instructions, I knew these lessons would not only serve me in performing magic initiations but also in my daily life, my work and my relations with others. El Anciano emphasized the difference between working with the right and the left hand when using the crystals. The left hand is the receiver of energy, while the right is the one that transmits energy.

When the purpose of a procedure was to emit energy (for example, rituals dealing with physical forms), I was to grasp the crystal in my right hand with its point away from my wrist. But, in situations where much more force was necessary, the crystal was to be held in my left hand (the energy receiver), pointing toward the wrist. If I wanted to both receive and send energy simultaneously, I was to take a crystal in each hand, thus maintaining an equilibrium between capturing and emitting energy.

El Anciano advised me to avoid holding the crystal for too many hours at a time in the same hand. When the crystal is in the right hand, a great moving force is acquired, but exhaustion can occur due to the large amount of energy expended. If this happens, it may be necessary to hold a crystal in the left hand for several days to recover. Likewise, keeping a crystal for too long in the receptive hand can disrupt one's energy field, causing insomnia due to the quantity of ac-cumulated energy. The best suggestion, then, is to alternate hands during the day whenever holding a crystal.

Self-protection is necessary during the prac-tice of magic. Most important is to establish a protective magic circle in a designated spot in the house. With this purpose in mind, I boiled 135 small crystals for half an hour and put them in the sun to dry, setting them down in a clockwise circle. The stones were placed very close together to create a tightly closed ring.

"If there exists even a small space between the crystals," El Anciano warned, "the exercise will not be successful. Step into the circle several times a day, staying for five or ten minutes. This should be sufficient to protect your aura from any negativity." He recommended that I do this daily, any time of the day.

Another kind of circle consists of crystals placed parallel to each other, facing inward without touching. In this position, they absorb the energy of the crystals in the interior of the earth and send it out their point toward the center of the circle where the person is standing. In this way, I received such an intense charge of cosmic energy that I was able to receive visions, perform magic ceremonies and take astral voyages.

Even El Anciano would always step into one of these circles before performing a ritual. When he left the circle, he was sweating and trembling, his eyes bleary from the tremendous force of the energy.

One can make an even simpler protective circle by using a single crystal (usually your personal one) in the right hand pointed away from the wrist. At the same time, mentally trace a red circle, as if it were a circle of fire, around yourself.

You can also envision the circle while standing upright, without moving, and making a circular motion around yourself with your arm. In this case, hold the crystal in the middle, leaving free its top and bottom ends.

"Anyone thus making a personal circle is protected against wounds," El Anciano assured me, "and nothing bad can reach him."

In order to strengthen the protective atmosphere, one can repeat this technique whenever necessary. According to El Anciano, the fire circle needs to be imagined not only around the person but also around one's personal effects, bed, work place and any loved ones who may be vulnerable.

The importance of pentagrams as self-protection was also revealed to me. I found out that the five-pointed star is a symbol of power and all invisible entities respect it greatly. El Anciano taught me to visualize and to draw red pentagrams behind doors of the house when I wanted to stop anyone from entering that area. As a general rule, the star must be drawn from the bottom right point with the single point on top. It can be traced on the chest, on the head or on anything or anyone that you wish to protect.

Each person has to discover his own unique symbol. Then, in order to protect yourself, thwart adversarial obstacles or obtain whatever you desire, the symbol should be traced in the air using the point of a crystal.

I learned several techniques to balance myself and promote positive feelings.

> *To begin, use a compass to find the four cardinal directions of the site you want to protect. Then place a crystal, facing upward, at each of the four directions. This sets up a positive energy field and prevents outside negative influences from taking hold. You cleanse the environment by tracing pentagrams of fire in the air with the crystal or by placing a crystal behind the door to any room you want to protect. This procedure should always be used before beginning any magic ritual.*

"Although invisible beings have the power to pass through walls, they only enter through doors or windows. To protect yourself, place a crystal behind any such opening to prohibit negative beings from entering, thus only positive ones will be able to pass through."

For this reason, I was told to put crystals in all the rooms, hanging them from the door frames by a piece of wire. This is similar to the miners' practice of hanging garlic in underground tunnels to protect them from bad spirits or vampires.

Personal exorcisms can also be performed on people with good results by using crystals. This is especially helpful when someone has been in a negative situation for a long time or feels the presence of heavy spirits. For example, one can achieve a "discharge" by touching the point of a crystal to various parts of the body, then moving it rapidly away from the body as if expelling something. This action should be repeated several times, progressing from the toes to the head.

In short, I was able to experiment with extremely positive vibrations and the enormous magnetic power of the crystallized gods. These exercises were practiced where the positive energies were so strong that I felt light enough to fly. El Anciano reinforced the need to put crystals throughout the house, placing them in every corner, revitalizing and adding more positive energy. "Even food and water become more powerfully energized when they come in contact with crystals," he said.

In fact, El Anciano had placed crystals throughout his orchard and even secured one in the water bottle from which we constantly drank. Our bodies were continually revived by their vibrations as the water would pass throughout our system by way of the blood stream. In the same way, homeopathic medicines multiply in their curative power when a small crystal is placed in their containers. This applies to beauty products as well. When the revitalizing energy comes in contact with the creams, lotions or shampoo, their positive chemical effects are accelerated.

I learned three ways to use the crystals to fight all kinds of pain, from paralysis to headaches. The first method is to place a crystal directly on the affected area so that its curative power will slowly nullify the pain. The second method involves holding a crystal between the index finger and thumb. Then with the crystal's base resting on the painful area, the point is rotated outward. In the last method, the crystal is pointed at the affected area and rotated quickly as if pulling something away from the pain. When performing these three methods, one must maintain positive thoughts to maximize the vibrations of the crystal.

El Anciano taught me not just to cure pain,

but also to combat what causes it. When the Chakras are unbalanced, they can malfunction by becoming either weak or overloaded. A crystal, held so that both ends are free, can get to the source of the pain by either transmitting energy to or expelling energy from the Chakras. In much the same way, crystals can be held over the genital area to stimulate or calm the sexual Chakra.

Concerning paranormal phenomena, crystals have an extraordinary influence. If a crystal is placed between the eyes, the third eye will stimulate paranormal activity that may have been dormant, such as a sixth sense, telepathy, clairvoyance, premonitions, magnetic cures or astral visions.

It was through El Anciano that I came to know about the Counseling Crystal, a point about 15 or 20 centimeters long whose base keeps it upright in a straight vertical line. This crystal, with permission from the universe, is able to reveal anything that one wishes to know from the past to the present and into the future. It is a magic stone with unlimited power that carries the deepest forces of Creation. This channel for communication with invisible beings is able to immediately and without fail manifest one's bidding. The name called forth by the crystal is equal to the sound of an enormous drum. It acts like a gigantic sun, drawing energy and astral beings from all levels.

According to El Anciano, there are several such stones in existence, and they have been responsible for the success of great historical figures. Plato, for example, through the use of his crystal was able to advance the knowledge of civilization beyond its time. Alexander the Great conquered the world guided by Aristotle, who possessed great tactical knowledge obtained through his crystals. And Napoleon was always led by the secret magic of his Counseling Crystal. However, Napoleon ultimately failed when he tried to destroy his crystal so that he could absorb its power and knowledge and thus transform himself into a living incarnation of the stone.

"The Counseling Crystal is very precious and must be treated with care. It should be covered with the finest black velvet, cleaned daily with a damp cloth, given small offerings of fire and used exclusively for magic rituals. It is only by doing all this that we can have enough dignity to own such a crystal and hope it will respond to our desires.

"We reach the crystal by means of the elements—burying it, in the case of the Earth;

tossing it into the sea, in the case of the Water; leaving it near a continually lit fire; or transforming it into vapor and spreading it through the air. Thus, we activate a type of soul-mate action within the kingdom, mobilizing all the elements so they can finish the program, accomplishing it in an infallible manner."

I noticed that before beginning a ritual, El Anciano moistened with saliva the crystals he planned to use. He explained that saliva carries the magic power of the blood, which gives us our vitality. Saliva, sweat or any other human liquid contains essential ectoplasm that will be absorbed by the crystal to unlock its power.

In one ritual, El Anciano first washed a crystal in water and then passed it nine times over a flame. Then, he held it in his mouth, supported by his right hand, for over an hour with the top pointed toward his throat. Meanwhile, the saliva that he swallowed had been energized by the crystal and his whole body received a powerful charge. In order to help the crystal transmit this energy, his thoughts during this time remained positive. This technique can also be used to help focus the mind before meditation.

"We should never accept a crystal as a gift, unless it comes from a well-trusted friend," cautioned El Anciano. "A crystal is very impressionable and can be pre-programmed. If it has been magnetized with unconscious energy, you will feel obligated, without knowing why, to follow the person's instructions. This could lead to your own self-destruction. Likewise, anyone so affecting a crystal should not forget the absolute consequences of the boomerang effect."

During my first weeks with the wise old one, I experimented by programming crystals in the Fire Ritual.

> *The Fire Ritual is always conducted during the last phase of the moon. Three candles are placed in the form of a triangle on top of a table covered with a black cloth. After laying out these items, the ritual can be carried out freely, letting your intuition be your guide. Usually, the crystal is put in an open steel container with alcohol. The container is surrounded by nine crystals facing inward.*
>
> *The liquid is lit, and after it begins to burn, project what you desire into the center of the crystal. When the liquid evaporates, pour more in and relight it two more times for a total of three times. When the ritual is over, the crystal should be cleaned carefully with a dry cloth and wrapped in a piece of black velvet.*

This ritual can be repeated as many times as necessary to achieve results, but keep in mind that success always depends on your capacity for concentration during the process. There is one important detail to remember: the candles should be put out with saliva and not blown out. Candles that are blown out acquire a magic force and can reject the presence of the persons around them. Putting them out with saliva, however, increases their power and permits them to be more forceful during the program.

The Fire Ritual functions with surprising efficiency in dissolving any kind of problem.

> *After performing the basic procedure, project the issue into the crystal and visualize its destruction. Then smash the crystal into bits with a hammer, tossing the fragments into running water.*

After participating in the Fire Ritual, I became even more aware of the enormous power within the crystals. They give power to projected thoughts and intensify the process of the ritual, using all of its elements to transform into physical reality what the magic desires. I now find it unquestionable that all human activities can be helped by the crystals . . . from the magic art of tarot cards, astrology and numerology to the work of doctors, psychologists, teachers or laborers.

One of the aspects that most impressed me was the force of the crystals, not only in the magic rituals but also in everyday life. The homage to the crystals El Anciano taught me was a clear example. Besides completing this homage before any particular practice, I was to offer such a greeting in the morning to the gods. The homage consists of a group of signs performed in order to remain united with the crystallized gods, thus attracting the cosmic energy they possess.

> *The homage or greeting is initiated with both hands placed in front of the chest, holding a crystal, preferably a long one pointed upwards. The thumbs should rest on the lower chest, over the solar plexus, with the fingers clasped around the crystal forming a triangle. Facing toward the west, close your eyes and concentrate on the harmony of the universe. Imagine the cosmos as a series of infinite shapes created by and emanating from the crystal until these forms surround the earth, our solar system and all the galaxies.*

> *Slowly, begin to unite your fingers and thumbs, wrapping them around the crystal, lifting it up above your head. Go up until your elbows come out to the side of your head. Your arms will form a diamond shape. Keep your arms in a vertical position in relation to the spinal column. Then, lowering your arms, trace the shape of a diamond with the crystal over your head, collar, chest, stomach and genital regions. Next, trace*

this shape in the palm of each hand. Finally, lift your arms again, forming once more a large diamond shape with arms overhead. With this homage, the energy of all the crystals in the world converge into your being.

During my initiation, I also learned how to program my body with the Elixir of Long Life. El Anciano said this elixir enabled him to run as fast as a child and have the sexual energy of a young man, even though he was of advanced age. He never confessed his age, assuring me I wouldn't believe him.

To obtain the elixir, place nine crystal points in a steel container filled with alcohol and heat them every day for several minutes over the span of three months. This ritual, begun at the new moon, can be performed either in the morning or evening but must be done every day without fail. At the end of each session, after the fire has gone out, place a fully-opened red rose next to the crystals. Clean the crystals and the container with a dry cloth and place in a piece of red velvet. After three months, put the nine crystals into a well-cleaned, empty wine bottle. It doesn't matter if the crystals remain clear or have developed cracks due to the repeated heating. Next, fill the wine bottle with red wine and cover with a red cloth.

I drank a glass of this elixir every day, either in

the morning or evening, without fail. This practice, attributed to Cagliostro and others, was used by alchemists in the Middle Ages who wanted to live long enough to convert lead into gold. This was also the secret formula many of the Popes drank in order to live a long, vibrant life.

A different type of ritual El Anciano taught me was one he used to achieve a particular desire. He placed on top of a table the number of candles equal to the letters in his name. On each candle, carved with a sharp crystal, was the corresponding letter written in Hebrew. Under each candle, a crystal was placed. After the candles had burned down to the crystals, they were gathered up, put in a pot of boiling water and left for awhile. Then the water and crystals were thrown into the stream.

The most fascinating of all the techniques was the Ritual of Evocation. In this ceremony, an invisible being could be called forth to materialize, and this being would remain in contact with the person who had called it. This was performed through a process of using fire and crystals; the crystal was used to communicate with the being and, unable to resist this call, it would appear. El Anciano would mix blood and saliva with the designated crystal, use magic words and wave a crystal rod with a very sharp point. He always carried out this ritual in the open air and at night. This was to be my next experience!

CHAPTER IX

THE EVOCATION

The possibility of summoning beings through the power of the crystals made a great impression on me. The image of a spirit I'd dreamed about as a child, and which had stayed with me until adolescence, appeared in my subconscious. Misraím, as I had named him, once played a crucial part in my life. Now I could have the opportunity to meet him again!

"Anciano, I'd like to perform the Ritual of Evocation, to see again a spirit from my child-hood."

El Anciano countered brusquely, "Misraím?"

He knew of its existence! But how could he? I'd never spoken about it to anyone! Well, probably the crystals had told him.

"Yes," I answered with surprise.

"Rama, Misraím cannot be summoned."

"Why not? Haven't you told me that with this ritual anyone can be called forth?"

"In reality, yes, but with some it is best to not try. And Misraím is precisely one of them. He appeared in your life at the right moment and if the universe made him disappear it was because the time had arrived for that to happen."

"But, I'd very much like to see him! It is what I have wished for many years!"

"Listen, Rama, Misraím is not exactly who you think he is. He appeared to you in the most appropriate form for that time, but that doesn't mean he is really just as you saw him. What happened happened, and you should forget about it!"

I could not understand why El Anciano was opposing my request so strongly, and I kept on pressuring him. "But, I'd like to see him again and chat awhile."

"Misraím is now in another place carrying out another mission. He shouldn't be bothered . . . and, besides, he could harm you."

"Hurt me? How? He was my best friend! Even so, I'd like to see him even if only for a short while."

"Well, if you want this so much, maybe it's because you're supposed to see him. But, pay attention, Rama. Misraím is not an angel; he's a demon."

"What do you mean by that? I know him. I know what he is!"

"You should not interpret what I've told you in a Christian context. Within the powers of the universe, there exists more than what are called angels and devils. In reality, there are many different forces with various functions."

"Then Misraím is a demon?" I asked, feeling fear rise up in my body.

"I just told you some of those we call demons are not really devils." El Anciano answered, a bit perturbed. "The demons are as good as the angels. They help us just as much and there is hardly any difference in the way they work. Misraím can continue to be your friend. It's only that now he lives for a different mission."

"I can't believe a meeting with Misraím would be risky or dangerous. I'm totally sure I want to summon him."

"I have a feeling the universe does want you to see him again. Nevertheless, in order to summon him correctly, you will need to obligate him to come, and for this I will have to help you. If something should happen to you while you are here with me, it will be spread about that I practice black magic," he murmured sarcastically.

In a few days, El Anciano informed me the time had arrived. It was a full moon and we got to the clearing in the forest a little before nine. We performed the Ritual of Homage, we exorcised the area and we purified our auras. Over our nude bodies we wore black capes, bordered with sacred symbols. He ordered me to be silent, to concentrate and observe his movements. El Anciano started to consecrate the site and all the utensils he carried: the gold medallion, the crystal rod, the sword, the incense and the coal. He chanted words and traced signs in the air with a long, fine crystal point which he held in his right hand. From his velvet pouch he took out small crystals and scattered them about the forest, forming a series of sacred symbols on the ground. From another purse he removed yellow powder and threw it into the air. From time to time he stopped, closed his eyes and remained very still as if meditating or gathering his intuition. He walked several times around the clearing, spelling out names, grasping his crystal point and observing the position of the stars in the sky. The only light was the brilliant moon-shine.

Once he had finished with these preparations,

El Anciano went to the center of the clearing and, with his crystal, slowly drew two circles side by side. He then wrote names and signs between them. Inside the circles he traced more symbols, including one right in the center where he was to stand, and one in front where I was to sit. We were to remain in place during the entire ceremony.

Three meters away from one circle, he drew a triangle on the ground. Around it he wrote words, and in the center he placed the point of a very large crystal. Finally, between the circle and the triangle, he placed a steel container with alcohol, herbs and flowers. He then moistened with saliva not only the lines of the triangle but also the names that were written in its center.

At exactly nine o'clock, El Anciano began the ritual. Standing and grasping the sword in which an embedded crystal reflected the moonlight, he addressed the hierarchy of beings to which Misraím belonged, calling out the names of each legion and its leader. He lit a rod and threw it into the container, lighting the flowers and herbs. Facing the triangle, he raised his arms and head upward while he pronounced other names. Sometimes he stomped violently on the ground as he shouted a series of strange words. In a short while, his gestures and pronouncements became extremely vehement.

Something was in the air, and I began to feel distinct, unexplainable impressions that grew stronger and stronger. Trying to understand them, I realized I had experienced these sensations before! My childhood experiences of Misraím returned—the same smell, the same light wind, the same vibrations and the same sounds. A breeze came up, the fire crackled and the insects around us were obviously disoriented.

What happened next was so rapid and surprising, I can barely describe it in words. I wasn't seeing Misraím in the literal sense, but the sensation of his presence vibrated with such clarity, it was as if I were seeing him with my own eyes. I was in such a heightened state that I felt I might see him at any moment.

Hovering above the triangle in a ray of light, Misraím appeared; it was the same sweet, yet almost witch-like, being as before. He smiled smoothly and asked me to come nearer. "Hello! You remember me, don't you? Come here, come!"

He spoke to me jumping up and down and laughing, but somewhat hastily as if he were in a hurry. I felt so emotional to see him again that I started toward him, but El Anciano grabbed my arm and made me sit back down. He continued calling, "Come here, come here. It's me, don't you remember?"

What memories! "Come here, loved one! How happy I am to see you!" It was really Misraím. Nothing had changed! And, I was so taken with seeing him I didn't think to look at El Anciano. I was so happy that I arose again, but El Anciano hit my arm and I had to sit down.

"Come, come!" Misraím begged. I was so deceived by the brightness, I didn't even feel the blows El Anciano had given my arm. Suddenly, he raised his sword shouting out loudly and imperiously. The brightness diffused and in Misraím's place was a grotesque and repugnant creature, his eyes and snout fixed on me. He was ready to devour me. He appeared hungry, thirsty and ready to kill. I was nearly paralyzed with fright. Then El Anciano gave another shout and raised his sword, threatening the creature so that the apparition left. It was over.

It took some time for me to realize and understand the paradox that had taken place. Besides the horrible fear this creature had awakened in me, I couldn't help feeling sympathy for it. When I knew Misraím, he had been totally different. We had been very good friends. I started to wonder about "good" and "evil," about the forces that, in fact, exist in the universe. Finally, I came to the same conclusion as El Anciano.

"The universe is all knowing and one has to respect everything that it does. When it gives or takes something away, that is what is best for us. Magic rituals should never be done merely on a whim or for curiosity."

"And what would have happened if I had left the circle?" I asked curiously.

"I don't know. Because your will was very strong, perhaps the universe would have protected you, or perhaps it would have allowed Misraím to enter your aura and he would have succeeded in taking you with him. I kept you from leaving the circle—as long as you were in it, you were under my protection. But let this be a lesson; once you leave here you will be on your own with the gods, and you will have to learn how to care for yourself."

"I never would have imagined that Misraím could have been that creature. I hope he doesn't return!"

"Listen, Rama, I repeat that there are neither angels nor devils. You should always explore questions that carry some kind of duality: good and evil, pretty and ugly, true and false. If you think about it, everything is relative. Duality is only the invention of organized religions to make humanity believe the Church and its representatives are correct and those who oppose them are wrong. When you understand how deceptive the world is, that the universe is infinite and everything changes and transforms itself continuously, then you will understand that the concept of good and evil is totally absurd. Everything forms a part of a whole, everything is equal and we all behave the same. In the universe, all is good.

"The separation of black magic, thought to be practiced purely for personal reasons, and white magic, said to be done for good, is a mistake. Ultimately, any will strong enough coming from deep in the soul does not represent a personal desire but rather one from the universe, which is using the soul to achieve it.

"You know that I feel a great admiration for the old religion and for the primitive rituals practiced by our ancestors who were in direct contact with nature, without dogmas and rules. Early peoples showed an attitude of respect for life that organized religions have rejected. They were spontaneous and practiced a simple magic, natural and without intellectualization.

"Magic is inherent in life, a part of everything. The universe is pure magic. Even animals practice it when they hypnotize their prey before an attack. Everything is pure magic: the atoms that make up matter, the transformation of cotton into cloth, birth and death. Those who birth ten children and those who perform magic ceremonies are doing virtually the same thing."

I listened attentively to his words as he went on with great enthusiasm. "The advantage of putting crystals in everything comes from the fact that they *are* the universe, in the form of crystallized gods. They attract the best energy that exists while powerfully helping the evolution of a person and all who surround him. The simple act of having and using a crystal represents the highest honor to the universe. Through the crystals, Rama, we can understand that life is a gift. Men are in paradise and they don't realize it. Drinking wine, dancing, loving women—all these show a respect for life, a homage to the Fire Spirit which symbolizes joy and happiness."

Chapter X

The Storm and the Village

One day El Anciano, looking worried, declared: "A strong storm is on its way. I want you to follow the river until you reach the first village you see. Warn the villagers that heavy rains are due and the river will flood violently."

"I don't remember having passed any village nearby. Where is it?"

"A few kilometers away. If you go down river, you'll reach it in two days."

"Anciano, I have always lived in a city. Why don't you ask the foreman to go? I don't know the forests. Where am I going to sleep? What will I do?"

"It will be a good experience for you, Rama.

The real jungle is not in the forest, it's in the city. Facing the wilderness will increase your capacity to fight against the jungles of men."

The following morning before dawn, I began the trip with great anxiety. All I carried was a knapsack El Anciano had prepared for me. This contained a wrap, matches, some cans of sardines and a can opener. In my left hand, I held a crystal to guide me through the forest and increase my intuition. On my mind, I impressed the last words of El Anciano: "When you arrive at the village, ask for a German named Bergen who lives in a small blue house and sells groceries. Tell him about the flood and then return here."

I headed out following the banks of the stream until the brambles were too thick. There I was forced to go into the middle of the stream in order to walk. Sometimes the water reached up to my waist and the bed of the stream was covered with sharp stones, making my progress slow and torturous. Mosquitoes pursued me with a vengeance, and I lost all perspective. Wild thoughts began to rush through my mind.

"What if a jaguar comes to drink water just as I am passing by? What if I meet up with a porcupine? What if a scorpion appears? There are many creatures about! Well, I could climb up a tree or, better yet, attack them with a stick. But what about the crystal? I could throw it into the face of any creature that dares come near me! What will happen to me?" I began to calm down, wondering if the crystal or El Anciano could be teasing me by putting such thoughts in my mind.

Now that my fears had quieted, I continued on my journey. Sensing my surroundings again, I noticed that much time had passed. What had once been just a stream was soon a large river. It was impossible to travel along the riverbed so I negotiated the overgrown banks once more. My progress was slow and, near exhaustion, I finally had to stop.

Gathering some branches and leaves to make a fire, I proceeded to set up camp. Much to my dismay, I couldn't start the fire. Reluctantly, I ate the canned sardines with my hands then, sticky and wet, lay down in a small secluded spot. I became so frightened by the sounds of the jungle that it was impossible to sleep. Huddled in terror, I heard a noise which almost froze my blood. The sound was so close I was certain it was a jaguar!

I couldn't just lie there. So even though I was scared out of my wits, I got up to investigate. I crept a ways into the forest, stopped and then stood still, hoping not to be seen. After a few minutes, it dawned on me there was no jaguar, and I turned back toward camp embarrassed and ashamed.

I realized that if El Anciano had sent me on this mission, he surely must have had a reason. His faith in me demanded I come to peace with my fears and not run from each new experience. With regained pride and conviction, I felt ready to face all that might happen. El Anciano had always said, "Fire ought to inspire water." To me this meant that my spirit could conquer any hostile environment. With only this thought in mind, finally I was able to sleep.

I awakened very early in the morning covered by mosquito bites. But even though I hadn't slept well, I was filled with energy and eager to get started. With new-found enthusiasm, I walked along the river for hours without eating. My surroundings somehow lost their menacing look, and I began to appreciate the wonder of the forest. I remembered a book I had read before meeting El Anciano. On the cover, Narcissus was studying his reflection in a lake. Just like Narcissus, I saw Nature reflecting her own beauty. All my fears were transformed into happy chants as I hurried to the village.

The sun was almost straight above when I detected a few wooden houses ahead on the banks of the river. There were two or three children at play and a small bridge on which a boy was calmly fishing. "At last, I've arrived! I am in the village!" I sang to myself. I rushed forward, hoping to speak with the children. But when they saw me, the children stopped playing, the boy lifted up his fishing pole and they all fled. As I advanced into the village, all the women gathered their young ones, putting them in the houses and closing the doors.

"Do I look so bad?" I thought. "All I need is for them to kill me!" I walked slowly toward an old man who crossed himself and ran inside. "This seems like a village from the Middle Ages! I can't believe it!"

Up ahead, I found the blue house El Anciano had told me about. It was open and in the doorway was a very fat man with blond hair and blue eyes who was watching me carefully.

"I am looking for Mr. Bergen," I said.

"That's me," he responded dryly.

"May I come in?"

"Enter!"

The house appeared to be a kind of store, and the German man invited me to sit down next to a counter. "What do you want?" he asked.

"I bring a message from El Anciano," I replied, suddenly feeling a bit shy.

"Yes, I know him. How is he?"

"Very well."

I proceeded to give him the message, just as El Anciano had directed. Mr. Bergen closed the store, heated up a plate of food and served us both a shot of aguardiente. He seemed happy to see me. He was quite talkative, becoming even more animated as he drank more liquor.

"You can sleep here today but tomorrow you should leave before dawn. Don't talk with anyone in town—they are dimwitted and ignorant. El Anciano helps them, but they remain afraid of him. He shouldn't help people so backward and superstitious."

"But how do they know I'm from his house?"

"They saw you come down river and knew it could only have been from El Anciano's. They think he is evil . . . and don't understand anything. I've lived here only a short time and will soon leave; I don't like stupid people."

We went on talking and drinking. "It's been a long time since I've seen El Anciano. I'd like to visit him. He is a very great man, very great!"

I saw a small scale on the counter and surmised that, besides selling groceries and utensils, the German probably also bought gold from prospectors in the region. In fact, this was likely the main reason he was here.

There was one main road through the village. During the rest of the day only one car passed by, stopping briefly at the store to purchase a few things. Later in the evening, he showed me where I could sleep, but before saying good night, he asked, "What is the name El Anciano gave you?"

"Rama."

"Aha. I understand! I get it!"

The following morning before dawn, Mr. Bergen bade me farewell. The return journey would be even more difficult and take more time because most of the way was uphill. I walked for two days, stopping only to eat and sleep when it was so dark I could no longer continue.

At the close of the third day, I finally came to the place where the stream parted from the river. I lay down to sleep when a light rain began to fall. The storm grew stronger and stronger, waking me at about midnight. I was drenched and could tell that the river was flooding. I knew I had to get back to the house before the weather worsened; so in spite of the darkness, I started to walk again.

The stream's current was so swift I could barely walk, even along the shore, and had to keep passing through the dense foliage on higher ground. The heavy rain was now a torrent and the creek started to flood the forest. Feeling desperate, I became totally disoriented. Dawn came and I had not succeeded in getting out of the forest. I was filthy, limping and bleeding. Ignoring the sharp thorns of the brush, I pushed on searching for a familiar sign by which I could orient myself.

Suddenly, I remembered the crystal. Sitting down on the earth, I squeezed this welcome friend between my palms, concentrating deeply on how to find the return path. After a few moments, I went on my way, feeling more sure of myself. Just a few hours later, I arrived at the foot of the hill leading to El Anciano's. I climbed the slope in the heavy rain and at last knocked on the door. I was beginning to think I would never reach the house. El Anciano, smiling and wrapped in a shawl, opened the door. "I thought you were never going to return."

"Me, too."

Soaked, I entered the room and hovered near the fire. The rain continued for five more days. "Anciano, how did you know there would be such a storm?"

"There's a place where I often go, and in that spot I became aware of what was going to happen," he explained.

"Is that where you go when you disappear all day?"

"Yes, sometimes."

"Could I go there one day?"

"Yes, when it dries out. It must be inundated now."

"Flooded?"

"Yes, it's a cavern."

CHAPTER XI

REVELATIONS OF THE HISTORY
BEFORE THE KINGDOM OF AGHARTA

A few days after the storm, El Anciano woke me up one morning. He was laughing and his eyes were full of mischief. "The people of the village may fear me, but at least they are thankful. Go look on the river bank."

My curiosity peaked, I ran down the hill to see what they had brought him. I could scarcely believe my eyes! There were animals, cages, furniture, pots, barrels of wine, a few lamps, clothes and even a pig. It was incredible! This was their gratitude for having been warned of the flood. No doubt it had taken the villagers many days to bring all the presents here.

When I returned to the clearing, I found El Anciano seated near the house, puffing happily on his new pipe. Somewhat sarcastically, I commented, "In light of all they've brought you, surely you'll be able to save much of it."

"So it is," he answered without stopping his smoking, obviously very satisfied with life.

El Anciano had promised to take me to the mysterious place where he spent so many hours, and he was good to his word. A few days later, we set off early in the morning. After walking through the hills for a couple of hours, we crossed a small valley. We walked up to a large rock formation where there were signs of former excavations and an abandoned mine.

I saw a narrow passage in the rocks that must have been part of the old mine. We took this passage which led us into the mountain and down into the earth. We had walked several yards when we came upon some old boards covering the entrance to another tunnel. El Anciano lit a candle that he was carrying in his pocket and gave it to me. Tearing away the boards from the entrance, he told me to enter the passage.

As I walked into what turned out to be a cavern, my eye caught something shiny and colorful. I moved toward the sparkling light on the other side of the grotto and saw a massive deposit of crystals. Looking up, I saw that the cave, the walls and the ceiling were all dotted with them. I knew this type of deposit existed, but I had never been in one.

When I held the candle at a certain angle, the flame reflected itself in thousands of small crystal points jutting from the walls. Ecstatic, I felt as if I were alone in space, surrounded by thousands of brilliant stars under a magnificent sky.

At the slightest movement of the flame, the crystals became a kaleidoscope of colors and light. I saw an infinite number of constellations illumined before my eyes. And, when I felt the soft caress of a breeze on my face, I had the sensation I was not standing but flying through space on a puff of air. In the middle of the cavern was an empty spot on the floor. "Yes," I thought, "that must be where the gods are."

Pulling me out of my reverie, El Anciano's voice beckoned. "Here is where I spend the hours alone, looking at the stars and talking to the Fire Spirit."

El Anciano and I returned to the cavern again and again. We would spend our evenings in the cave talking and performing rituals. It was during these nights that he first mentioned the History. This is what had been revealed to him through the stars and his conversations with the gods.

After the Great Sin, there existed three other civilizations before ours—Hyperborei, Lemuria and Atlantis. These civilizations were superior to ours and all of their knowledge was based on the power of the crystals.

Atlantis was El Anciano's main focus. In Atlantis, men were very different than they are now. They cultivated crystals, fire and the serpent in that Solar City. By using the crystals, they connected back to the time before the Great Sin and were able to restore their original powers. Crystals provided the energy for all of their needs. The stones were used in order to heal physical illness and even enabled the Atlanteans to fly.

But history was repeated and Atlantis began to get out of balance with the universe. A cosmic cataclysm destroyed it completely, leaving not a trace of what had been. Crystallized gods are all that remain of the lost paradise of these crystal cities. The few survivors scattered to regions we know today as America, Egypt and Tibet.

Realizing the destruction had occurred because of misusing the crystals, the survivors decided to remain silent forever about the hidden power of the stones and to keep their knowledge restricted to a select few initiates. Tibet is the place where the magic has been most impenetrable, and until recent times, it has remained there intact.

In Ancient Egypt, the knowledge was given to a few powerful people who used the magic for personal gain. This led to another great imbalance and the fall of this civilization as well. The knowledge eventually spread to the Mediterranean, and it was in Greece that a new place of refuge was found.

El Anciano told me that Sparta was able to conquer Athens only after a small fleet of ships brought crystals from North Africa. The Spartans used the crystals to weaken the Athenians and strengthen their own warriors. Crystal, Christos, Christ—the Greeks were the guardians of this knowledge and revealed it to the first Christians, the gnostic Valentiono. The origin of Christendom is found in the magic of the crystals, and it is a shame that this true Christianity has degenerated into what is practiced today.

During the Roman expeditions, a few initiates shared their knowledge with new disciples throughout pagan Europe. The disciples formed small groups and spread what they knew about the crystals to northern Ireland, Holland, Switzerland and Germany.

This wisdom survived violent religious persecutions, passing through a few secret societies during the Middle Ages. Many new disciples fled to Asia and North Africa, forming secret societies that remained hidden under the guise of Islam.

During the Middle Ages, there were two religious-military orders that promoted the knowledge—the Knights Templar and, to a lesser degree, the Teutonic Knights. Some of their major forts were constructed in the shape of a hexagon, which mathematically reflects the internal structure of a crystal.

After so much dispersion, the original wisdom became corrupted in Asia and North Africa. Meanwhile, in the New World, which until its discovery had remained fairly removed from "civilized" history, the descendants of Atlantis were able to maintain the original teachings through the indigenous people. The shamans, with their primitive magic cults, rejected the intellectualization of the Europeans and kept the old knowledge within the tribes. Thus, the ancient beliefs were transmitted principally by the Indians of North America, Mexico and Brazil.

Still, only a few "campesinos," the shaman and the highest priests of the indigenous cultures, had access to the knowledge. The teachings were transmitted orally, passed down to direct descendants only. In the Americas, crystal deposits were considered sacred ground, and any strangers who happened along were destined for death. Strife between tribes occurred not over the conquering of cultivated land but over possession of deposits of high quality crystals.

In the Americas, the land in Brazil was the most sought after by the initiated because over ninety percent of all existing crystals on Earth are there. The crystals create a strong electromagnetic field. It is for this reason that Brazil has always played a special role in the practice of magic.

Pedro Alvares Cabral is an example of a disciple whose mission it was to discover the "sacred ground" filled with crystals. There were many others who supposedly came looking for emeralds but whose main objective was to reveal the locations of the great crystal deposits.

One night in the cavern, El Anciano told me an amazing story of the past. "Over a long period of time, in the immense region of the Amazon, there existed a great civilization that practiced the cult of the crystallized gods, much as did the Atlanteans. Unfortunately, this settlement began to commit the same mistakes over again and so it was destroyed."

"Really, it seems incredible to me that the wisdom has been able to survive in spite of so much confusion," I said in wonder.

"The universe found the way to preserve it." I

suspected this was a story in itself.

"How?"

"Through the creation of an underground kingdom that maintained this tradition after the destruction of every great civilization, one that exists parallel to ours even today without mankind being aware of it."

I was truly astonished; I had never heard of such a thing! "Tell me more about this kingdom," I asked curiously.

"When a group of wise men from Atlantis foresaw the coming catastrophe, they gathered all their knowledge and disappeared into the depths of the earth, through secret entrances leading to the center of the planet. Earth is not solid; it is threaded with tunnels and galleries that unite all the continents. Safe in the interior, these wise men survived the catastrophe and organized a very advanced harmonious society. Intending to remain forever within the Earth, they built cities and created a perfect society.

"The center of this world was called Agharta and there they practiced the greatest Fire Rituals. Their leader was a human incarnation of the Fire Spirit, called either the King or Prince of the World.

"Underground, they continue to practice the cult of the fire, living from the power of the crystals. From inside the Earth, their influence extends to the events of the outer world. For example, in the past when they saw that the original knowledge was degenerating, they sent forth emissaries to reestablish their hold in society through new secret circles.

"Because the largest concentrations of crystals were in South America, the main tribes there found themselves doing quite well. The Incas as well as the Mayans were tribes of Agharta that surfaced in this way. Before the barbaric comedy of the Spaniards, they returned to the interior by the same passages from which they had come, sealing off the tunnels behind themselves. Many tribes considered hostile by the Spaniards knew of these entrances, having been left, in fact, to guard them."

"And this kingdom still exists, with people really living there?" I asked. I had to know everything; it was so intriguing. Either El Anciano was crazy or I was going to become involved in something really fantastic.

"Of course. They breathe the air that passes through all the underground galleries and harvest food that grows around their cities. The crystals give vitality and light to the crops."

"And can one reach this underground kingdom?"

"Many entrances exist but most are hidden; still others lead to mazes to confuse the curious. Some of the entrances are in the towns of Minas Gerais and Mato Grosso. One of the main ones is near San Lorenzo, and there is even one near here," he said.

"Near here? Have you been there?"

"I know where it is, but I have never entered."

"Have you had the opportunity to see some of the inhabitants of the Kingdom of Agharta?"

"Of course! Once or twice a year I get in touch with them through the crystals, and we meet at the entrances to the galleries."

Dying with curiosity, I implored him to tell me about everything. "And what do they look like? Are they like us?"

"No, they are much taller and their eyes are different."

"Do they speak our language?"

"No. They have their own language, but through the crystals one can communicate in any language."

"Could I see them, too?"

"Perhaps," answered El Anciano without showing any interest. It seemed as if he wanted to close the conversation. "If you are prepared for this and they want to see you, who knows?" And with that, he ended the discussion.

Chapter XII

A Trip Through the Stars

"Rama, the day is approaching when we will make a Great Fire Ceremony to celebrate the crystallized gods and their union with the Fire Spirit. We'll worship the memory of the Great Sin and pray that the forces of the fire will return the world to its previous state. You will be present on this sacred occasion. You'll see the Fire Spirit, and later he will go with you."

"Go? Where?"

"To your home."

I started to shake. I had not thought of going back home. The only thing I desired was to stay here.

"Are you surprised? I took away all your

worldly illusions so that, knowing about the invisible beings, you could return to your own life and contribute to the evolution of the universe. This is not Shangri-La. There is no Shangri-La in the real world, only a difficult life we must confront and not run away from. Nevertheless, you still lack something very important—the power of concentration without which little can be accomplished."

El Anciano invited me to sit next to him in front of a flower. He rested his hand lightly on a crystal that was on the ground, pointing the crystal and his fingers toward the flower. He stared at the flower without blinking, and in a few moments it withered. Then, rubbing his hands, he placed three crystals around the flower. Putting his hands on the ground in the direction of the flower, it came back to life! With wonder, I studied how the stem had righted itself and the flower, totally open, seemed even more beautiful than before.

"Now you do it," El Anciano ordered.

I repeated all his movements only with my eyes closed. I concentrated as hard as I could, until I could feel the veins in my forehead constricting. But when I opened my eyes, my enthusiasm had disappeared. Absolutely nothing had happened to the flower.

El Anciano reassured me: "Concentration is not a matter of muscular force, but of a straight and serene thought that perforates like a ray and acts as a trajectory. We are going to begin some exercises to develop your capacity to concentrate."

Following his instructions, I chose, by the method of irresistible attraction, some "virgin" crystals which I would later put to use. To give them strength, I passed them through the flame of a candle, first turning them upwards, then downwards, then horizontal. When I finished, I placed them carefully in a small velvet pouch.

The exercises required me to work by myself, seated within the outline of some very powerful star shapes. These stars were formed by placing crystals in the corners of an imaginary image on the ground. On top of a large boulder, near where the stream began, El Anciano showed me what to do.

"There are no formulas or specific rules in working with these stars. You freely engage your intuition and can even choose the kind of star with which to work," he said.

I decided to begin with a six-pointed star. I knew this was an exercise in self-control designed to balance the whole body. After I had laid out the pattern of the star with crystals, I sat down inside the stones, completely nude, with my hands on my knees, my back straight and my eyes closed. I put a crystal between each hand and knee, the one on the left with its point facing toward my wrist and the right-hand one in the opposite direction. The goal was to remain perfectly still, with my muscles tense and completely rigid, for a period of time. I was to do this exercise five minutes in the morning, afternoon and evening, gradually increasing the time to an hour.

After a couple of weeks, I was convinced I was doing well. But El Anciano told me my work was awful and that he scarcely noticed any force. So I had to begin again with a different exercise. This time I stood inside the star with a crystal balanced on top of my head. In the beginning, I felt terrible pain, but little by little I got used to doing it. When I was finally able to remain in this position for an hour, El Anciano thought I was ready for the second phase.

In this next exercise, I was to become totally aware of my breathing. Seated in the middle of a diamond shape and in the same position as before, I had to alternate closing one side of my nose while breathing in slowly through the other. Then, with both nostrils closed, I would hold a deep breath for as long as possible. I held a crystal in each hand, projecting my inhalations through the right hand and exhalations through the left. At the end of this exercise, I had established a very special sense of integration with the crystals. Through many hours of practice, I developed a sense of ecstasy in which I became one with the crystals.

The third phase was to take place inside a pentagon. In this exercise, I was supposed to develop my concentration to such a state that my only thought was the crystal. So, I began by sitting properly and breathing as I had learned in the second exercise. With my eyes closed, I visualized the crystal just as I had seen it—line for line, angle for angle. I had to keep focused until I was able to make it so real in my mind that I became the crystal.

After much practice, I could visualize it moving and revolving around my head. It was important to allow all my senses to interact with the crystal by feeling its touch on my skin, feeling it inside my mouth and hearing it grazing the ground. This exercise was the hardest for me. Holding the mental focus even for only a few seconds was very difficult. I discovered that dominating the physical form was easier than accomplishing the mental concentration.

Finally, El Anciano allowed me to move to

another level. "In order to master all parts of our being," he explained, "it is necessary to understand both the low and the high forms of religion. In reality, they are not opposites but rather different variations of the same hierarchy. A man and a woman, for example, have to join in order to have a child, and that child represents more than just their union. Through the third element, the joining, the child surpasses the parents. It is perfection that resides in this balance, or the creation of a new being."

El Anciano used a mysterious symbolism which, little by little, I was learning to decipher. He felt it was now necessary to teach me about the "inferiors," that is, the demon spirits called the Qliphots. These demons are members of the hierarchy and the opposite of the angels, whom I had been working with up to then. The exercises had to take place inside a triangle and with the same star shapes I had previously used, only now the stars were to be inverted.

Before beginning the new set of exercises, El Anciano drew a red triangle, point down, on my navel, repeating this act on each of the next several days. After I had completed the exercises to understand the demon world, he then drew another triangle overlapping the first one to form a hexagon on my body. This signified that I now was initiated into the wisdom of equilibrium; my former dual vision united, I could henceforth see the oneness in all things. All of the teachings became more coherent for me at this time.

After these last exercises were over, I was not surprised by anything El Anciano suggested or did. I knew I was getting closer to the experience of the Great Fire Ceremony. It was with this thought in mind that El Anciano announced, "It is important for you to pass a test to see if you have sufficient self-determination to participate in the Fire Ceremony."

"And what kind of test is this?" I asked.

"You must spend the night with multiple stars."

To pass the test, I had to be totally alone, from dusk to dawn, in an intricate pattern of stars all purposely mixed up. This chaotic pattern would attract a convergence of angels, demons, beings, spirits and gods from different dimensions and hierarchies, causing an "astral storm."

"In this way, it can be seen whether an initiate has enough fortitude to face any kind of magic exercise."

My only defense would be a circle traced among the stars and from which, under no circumstance, was I to leave. El Anciano had taught me so many things during these months that I felt totally secure and had no fear in facing the test.

In a clearing in the forest, he spent one whole afternoon preparing a multitude of stars full of signs and symbols and sacred names. A line traced on the ground designated those crystals that had been activated during the night in a bonfire. Some seeds and boiling water were thrown over the many stars and in the center of it all was a circle made of crystals. That same afternoon, around the stars and in the form of a triangle, El Anciano anchored three torches which were to remain burning all night.

At sunset he sent me to the clearing, "What do I have to do?"

"Simply remain seated in the middle of the circle."

Careful to not step on anything, which was quite difficult given the number of crystals and drawings, I entered the circle and sat down.

"Do not try to leave here on an astral voyage. If you do, you will never return. I'll come for you in the morning."

As soon as I sat down, I perceived something strange. I couldn't tell exactly what it was, but I felt intense vibrations that were much worse than those I'd encountered during my "meeting of the inferiors." It got dark and I became more and more agitated.

As the night wore on, an atmosphere of total terror began. It seemed as if a strange presence was behind my back; but whenever I turned around, it disappeared only to return somewhere else near my body. This happened so many times that I feared I was crazy. A strong gust of wind put out the torches, throwing them to the ground. Throughout the night, a great darkness covered the forest. I felt the presence of some "things" that flew near me. A deafening noise hurt my ears, and then I started to hear other sounds of anguish. I also seemed to see figures walking back and forth, but the worst was the sensation of changing gravitational forces. I felt as if my body were contracting, first between vertical walls, then horizontal ones. Other times I lost the sense of gravity altogether and had to cling to the ground fighting the expectation that a simple wind would come to blow me out of the circle.

The fear was constant, so I decided to practice the concentration exercises. This brought some relief, although only briefly because the forces around me were becoming stronger and stronger.

After what seemed like hours, the sensations began to diminish in intensity until they died down completely. The black mass lifted and the moon and the stars came out. Silence accompanied the night and everything returned to normal. "I've won!" my inner being shouted. "The nightmare has ended!" Encouraged by a sense of triumph, I calmly rested while admiring the moon.

However, it was not long until I heard a very peculiar sound. There was a hissing that was becoming clearer and clearer until at last I heard it with perfect clarity. "SSSSSS . . ." Immediately, I turned around and, in spite of the darkness, could see a few yards away a line on the ground moving toward me. It was a snake!

My first impulse was to run, but I remained where I was. My mind chanted El Anciano's instructions to not leave the circle for *any* reason. The snake was within the area of the stars, less than a yard away, fully stretched out and ready to spring. The only thing I could think of was to remain perfectly still, like in the hexagonal exercises, confronting the snake but without the slightest movement.

It came closer still and stayed by the side of the circle staring at me. Remembering my exercises, I was so still that I did not even blink. I fixed my eyes on it and tried not to show any fear, but the hissing grew louder. Out of the corner of my eye, I could see another snake coming closer! Its head was lifted and its scales shone under the reflection of the moon. Now there were two of them challenging me with cold, aggressive expressions. I was petrified!

Soon I heard the hiss of yet another cobra. More and more appeared until I was surrounded by dozens of writhing reptiles. They intertwined with each other, biting and fighting among themselves, looking for a way to penetrate the crystal circle.

"Are these real or imaginary serpents? Are they made up of matter or not?" I asked myself.

An idea occurred to me. If the crystals were still in place on the ground as before, then they were not matter because their movements would have displaced the stones. I tried to see through the dark, and it looked like the crystals were exactly where El Anciano had put them. The slithering forms might appear to be serpents, but they were not real matter. This thought calmed me. If the serpents were not matter, they could not enter the circle. The most curious thing of all was that after I concluded this thought, the snakes began to disappear as if by magic!

There still remained a strong presence in the atmosphere, however, and when I looked at a certain point in the forest, I saw yet another serpent. This one was different from the rest. It was much larger and thicker, almost gigantic. I didn't have any fear because I was projecting positive thoughts of security and wisdom to the creature. It stayed just outside the stars, which gave me some comfort. It had golden scales and was extremely beautiful. We looked at each other for a long time until the light of dawn appeared. The rising sun made the snake's scales glow more and more golden. As the day came, the creature started to disappear until it completely vanished. The sun now lit up the whole forest.

Not long after the snake had disappeared, El Anciano arrived to take me back home. Totally drained, I fell into bed and slept for three days. We spoke very little about my experiences, but it seemed as if he knew all that had happened. Although, regarding the golden serpent, he did explain it was one of the forms the Fire Spirit assumes, and the fact that it had appeared to me was confirmation I had passed the test.

"Now, indeed, Rama, you are ready for the Great Fire Ceremony."

Chapter XIII

Rituals of the Crystallized Gods

Ever since I started studying with El Anciano, it has been apparent that the gods were close by in the form of very pure stones. These were true deities worthy of being worshipped by individuals and groups. What I didn't know was that it was possible to worship them in two different ways: through magic, the path El Anciano had been teaching me, and by being a devoted believer.

"These two paths are very similar," El Anciano assured me, "although, the way in which they are practiced is different." He himself had chosen the way of magic, but he would show me the other path so I could chose for myself. In the first place, it is necessary to differentiate between the crystallized gods and their devotees. This distinction begins to disappear, however, through the practice of rituals. Eventually they become one and the same, as the god-filled crystals fill the heart of the believer.

The believer should construct an altar, a physical place, where he can practice his devotion. The size of the altar is not important—only the care with which it is built. As a general rule, it is best to use a special room reserved for the altar. If that is not possible, one can use any room of the house with the stipulation that when the altar is not in use it will be covered. Another possibility is to use a cabinet for the altar.

If the entire room is set aside, it should always remain closed and clean, and only the worshipper should have access to it. Nothing else should be stored there other than what is important in worship. The walls can be painted black, red, yellow or even white, according to personal preference and the type of energy needed for worship. The only light allowed in the room should be candlelight; however, an altar can be situated near a window where the crystals will receive the sun's energy when not in use.

The most important aspect is the altar itself. It will be the same shape whether it takes up a whole room or small cabinet. It can be made of wood or brick. All temples should be constructed

with three tiers and covered with a cloth, generally black or red so the crystals will reflect their utmost beauty and splendor.

On the first step of the altar are placed the most common crystals and those less frequently used. Special pouches of the stones and any jewelry containing crystals can also be placed on this bottom stair. The second step is where more important crystals are placed, those the worshipper feels great attachment to and ones used in magic ceremonies. Finally, the most important crystal is located on the third, top step. This is usually one with a long point, or one with a flat base whose points soar upwards. This is chief among all the crystals on the altar, the principle divinity, the one to whom the devotee will pay homage. If you possess a crystal geode, it is considered a god superior and should be put above all others.

You may also place there other kinds of stones such as tourmalines, citrines, topazes, emeralds, amethysts, opals or beryl. Just be careful not to put them on the same step as the divine crystals. They are best located on a separate table to the side of the temple.

Under the altar should be placed a foundation stone. This is a piece of gold or any significant personal object that is renounced and offered to the gods as a great gift of love from you. On the sides of the second step add two silver candle-

sticks to illuminate the gods. Hand drawn above the altar should be the symbol of the Sacred Tradition.

The more beautiful and harmonious the altar, the more the devotee's soul will embrace the sacred stones; thus, the worshipper's inner beauty will begin to shine forth as the altar is used. Do not hesitate to make your temple feel and look divine.

Everything that is used to adorn the temple should be exorcised and consecrated beforehand. Take each object and, with a crystal in your right hand, draw a pentagon and a diamond over it while chanting the words: "I deliver you to be consecrated by the divine crystals so you may receive the spirit emanating from them. And, in the name of the crystal gods, I free you from all influences you may have received before today."

After this exorcism, the crystals should be washed with soap and water and passed over the flame of a candle, one by one. Only then should you place them on the altar, with their points facing toward the interior of the temple. The ones that are supported by their own base may be left standing upright.

Once the altar is ready, the worshipper may begin the first ceremony of adoration. This should be started at nine o'clock, either in the morning or the evening. You should be completely alone and the temple closed to everyone *else. Kneeling down, gaze at the crystals and worship them from the very depths of your heart. Upon cherishing the divine universe, the most beautiful and heightened energies will come to you. After finishing this external worship, close your eyes and imagine the crystals inside of you, filling you with light. Finally, imagine yourself as a crystal and cherish yourself. At the beginning and end of each ceremony, always carry out the Homage to the Crystals. It is important to worship the stones in the nude so that you come to your temple pure and without ego or adornment.*

The Great Invocation should also be performed. This prayer is spoken using your intuition as your guide. The more you can allow your intuition to lead you, the more powerful will be the results. The invocation needs to have seven parts—seven paragraphs, seven sentences or seven words. There are seven messages to be transmitted to the gods during the Great Invocation—an Impression, a Judgment, a Memory, an Oration, a Conversation, a Conspiracy and a Madrigal. The first part is a message of humility; the second of fidelity; the third of dependence; the fourth of adoration; the fifth of confidence; the sixth of brotherhood and the seventh of passion.

Worshippers practice this ceremony in their own way, but it is best to follow this order: Preparation (by bathing or relaxing), the Great

Invocation, the Adoration (internal or external), the Sacrifice, the Contemplation-Meditation, moments of Communicating with the gods, the Homage and, finally, the Veneration.

"Besides being a location for the ceremonies, the altar can also be used for meditation," said El Anciano. "I repeat that it should always be kept impeccably clean and well cared for so that nothing profane will defile it. In his mind, the believer can hold the image of the temple so that it becomes a part of him, although the person will in actuality be separated by a physical space. Truly," he said devoutly, "the Temple should be kept alive inside the devotee as well as sacred on the outside."

El Anciano explained how reciting verse or poems in front of the altar is a form of homage to the crystallized gods, as is rising in the night to adore them. These practices will assist in developing a deep reverence. After a short while, the crystals will supply great strength and a connection to divine wisdom. The crystals and worshipper become as one, and this union will keep manifesting increased magical powers. The practitioner will eventually have the capacity to perform miracles and experience natural psychic powers such as telepathy and sexual potency. Finally, the believer will also be converted into a crystal god, in human form.

"This path requires a lot of persistence, Rama. Really, there are very few who manage to follow it, but those who do receive much in exchange." Ending his descriptions, El Anciano sent me off to build an altar to protect my crystals, regardless of whether or not I chose to become a worshipper.

CHAPTER XIV

THE VICTROLA RITUAL

One night as we were talking quietly in front of the fire, we heard a loud shout coming from just outside the house. Startled, El Anciano got up and I took protection within the magic circle. El Anciano opened the door to a man with an angry expression the likes of which I had never seen before.

"Yes, I have come," said the man defiantly as he stepped into the house. El Anciano remained silent as I looked on wide-eyed from behind the crystals. I didn't know what was happening, but I sensed something hostile between them. With a wave of his hand, El Anciano motioned me to leave the house and wait outside while they spoke. I obeyed and once outside started to pace around the house.

At first they spoke normally, and I could barely hear what they said. But then their voices got louder and louder, until the conversation became a violent quarrel.

"If you do this you'll ruin everything," shouted El Anciano.

"I told you I would do it. And I will!"

The man was accusing El Anciano of not directing the Brotherhood as he should and was threatening to make everything public, thus forcing El Anciano to resign his position. "You know you are totally alone and no one will follow you." The old man remained strong.

"I'll destroy everything, I swear it!"

"Get out of here. Get out of my house!"

The door opened and the man stood before me with a hateful expression. He shot a scornful look and, slamming the door, left by the same path on which he had arrived. I approached the house and asked if I could enter.

"I am very sad, Rama. Men continue to commit the same errors of the past and seem not to learn anything," he said looking dejected. "There isn't anything else to do but to stop him!"

"But, Anciano, didn't you teach me that man has a right to do all that his will dictates to him?"

"He is within his rights to do what he wants, but I also have the right to react to his threats and not agree with him. When two strong wills oppose each other, they must confront each other with force. He who wins is the most powerful reflecting the true will of the universe. The other will find himself in disharmony with Creation. I believe I am obligated to resort to a dangerous ritual in order to protect myself and the Brotherhood."

Towards midnight, El Anciano took out a leather valise from inside a trunk and opened it. Inside was a portable victrola, whose end was hooked up to a car battery. A candelabra hung from each side of the victrola. He put everything on top of a table that was covered with a black cloth.

Following this, he took a crystal I had never seen out of a small velvet pouch and secured it to a record with adhesive tape. Placing the record on the turntable, he started the victrola. The record went around and around so quickly that the crystal started to look like one big white circle. Seated in front of the table, El Anciano stared hard at the circle. Deep in concentration, he sat motionless except for his lips which were constantly moving although he didn't say anything.

In the past, he had spoken to me of the Ritual of the Victrola. I understood that its purpose was to create astral beings who would complete a planned mission. Projecting a mental image of a creature into the magnetic field generated by the rapidly moving crystal could make the entity come alive to obey the orders of its creator.

But El Anciano had always avoided this ritual because he considered it excessively dangerous. If the creature is able to control its own life and rebels against the creator, it becomes transformed into an "astral vampire," which can eventually absorb the master's energy and life. In spite of this, he was convinced he had to use this ritual against the man who had left his house threatening to destroy everything.

El Anciano remained immobile for a long while, staring at the revolving crystal on the record. The atmosphere was so tense that I decided to go outside and breathe some fresh air. Upon returning, I saw a scene that paralyzed me. El Anciano, standing in front of the victrola, was making a cut in the palm of his left hand. Then, squeezing the wound, he let drops of blood fall into a crystal chalice. He spread his blood over the victrola, the record and the center of the white circle.

An enormous black dog appeared. His back

was raised, his feet separated, his eyes filled with hate and his teeth bared as if ready to attack. He looked toward me, searching for the door. I darted away from it, and the strange creature shot out. In retrospect, it might not have been a dog, but this visible representation of energy had been without a doubt created by El Anciano, and it seemed absolutely real!

Exhausted by the superhuman effort he had just expended, El Anciano fell across the table. After a few moments, he began to rise up and slowly bandaged his hand. When our eyes finally met, his carried such a defeated expression that my heart ached for him.

Chapter XV

The Great Fire Ceremony

The morning had finally arrived! That day I would partake in the Great Fire Ceremony. It was a glorious day without a cloud in the sky. Everything felt different to me. All the crystals shone brilliantly, noticeably filled with a special light. And Mother Nature seemed full of happiness, as eager as was I.

After the incident that had caused El Anciano to perform the Ritual of the Victrola, we had used our time to prepare for the upcoming ceremony. It was to take place in one of the clearings near our place, one reached by a small twisting path through the forest which would confuse anyone who did not know it. Over the past weeks, I had helped El Anciano cut firewood and carry many crystals there. On special orders, I was to stay in my room to prepare myself for the ritual, repeating the exercises for concentration and posture. I fasted rigorously and rarely saw any sunshine.

This morning, El Anciano had come to my room to give me instructions. He told me about the Brotherhood of Fire of which I hadn't heard a thing since my encounter with Jonas in Minas Gerais, some months past. He explained that, besides himself, there were several other people in the group and they dedicated themselves to the study and practice of the crystal fire magic according to the Sacred Tradition. The majority of the people led a completely normal life in big cities, and nothing set them apart from others. This was an informal, disparate group, but they were constantly in contact with each other and met occasionally to perform certain ceremonies.

Actually, the Brotherhood wasn't just one group but many groups united by the leaders of each group. These teachers were joined in a hierarchy which echoed the spiral structure of the crystals. The members were free with no formal obligation to the Brotherhood other than to keep secret the most private practices.

According to El Anciano, the Great Fire

Ceremony is a major collective ritual of the Brotherhood and, thus, the various groups would be present for it. "This Ceremony pursues several ends, one of which is the direct contact of the Fire Spirit. This spirit, once invoked, appears before all the members, helping them to integrate the crystals and the spirit. During the ritual everyone mobilizes spiritual forces to work magically in causing the victory of fire over water, the evolution pathway of mankind. They call the most powerful forces of the universe and, above all, its own Fire Spirit called the Carrier of Light. "After the Ceremony," El Anciano assured me, "you will be a part of the Brotherhood of Fire."

As I was not yet a member, I couldn't actively participate in the ritual, but I could closely observe it. A few days before the ceremony, El Anciano showed me the circle of crystals from where I would watch the proceedings as well as the black tunic I would wear. I was warned that besides needing the protection of the circle, I must use a series of secret formulas he would give me throughout the ceremony. This would save me from the Fire Spirit who tries to kidnap and consume each new initiate. Even with protection I might feel strange things, but I was not to give in to my fear. Instead, I needed to remember all the exercises he had taught me, using them to sustain me through whatever might happen.

I had also taken a pledge to keep secret the identity of all who participated in the ceremony. The ritual was scheduled at nine in the evening, and when I heard the ring of the bell, I must leave my room, dressed only in the black tunic, and file silently in line with the others. After giving these instructions and explanations, El Anciano left me to be alone again.

In the afternoon I began to hear voices of people arriving, but I continued with my exercises of concentration. Once it started to get dark the voices quieted. At the designated hour, the signal sounded and I walked outside the house. There was a row of nine people wearing red robes and masks representing animals such as falcons, snakes, lambs and chickens. They stood straight in line with their hands over their chests, carrying a lit candle in the right hand and a small crystal held lower in the left hand. Later, I noticed that all wore a silver medallion around their necks with the symbol of the Sacred Tradition. El Anciano had a different kind of medallion, made of solid gold.

At the front of the row was a very short man, possibly a dwarf, who wore a falcon mask and carried a torch in his hand. Behind him came El

Anciano, gripping a glorious sword. He struck its blade on something that made a sound similar to the ringing of a bell. Everyone stopped moving and turned toward the front. He made another signal and the group walked silently on the dark path that led to the forest glen.

The sight of this phantom group with red shawls and animal faces was strange indeed. The dwarf's torch illuminated the way with the rest of the path lit by the group's candles. I followed closely behind, listening to the crackling of the torch and the sound of the many footsteps. The whole scene was surreal, like something medieval in the middle of a world dominated by science and technology.

After walking for half an hour, we left the path and arrived at the glen. In spite of being surrounded by trees, we could still see the hills. In the center of the clearing was a bonfire three yards high, made of branches, dry leaves, plants and enormous crystal geodes. Everyone fell out of line and gathered around El Anciano and the dwarf, who kept the middle of their circle focused with his torch. I went quickly toward my special circle to observe.

El Anciano gave another sign and nine Brothers of the Fire started the Homage to the Crystals. They made a diamond shape with their arms over their heads and chanted to the crystallized gods and the Fire Spirit. Meanwhile, El Anciano walked around each of them making signs over their bodies with his sword. After the chants, they went one by one to him, rubbing their masked faces on the left ear of the serpent mask El Anciano was wearing, and giving a kind of kiss or signal. Then he returned a sign of consent with his head.

A carafe was passed around from which everyone drank. El Anciano stood atop a rock and, with the sword embedded in the earth in front of him, began an exhortation. He spoke about the deeds of the Brotherhood and mentioned the most recent "transfer of knowledge" from the gods. He also talked about the Great Fall, of the apparent defeat of the Fire and of the victory Water would soon have. He spoke of the Great Fire Spirit—its glory, kingdom and power and its triumph over adverse forces. He also explained the desire of the crystal gods to return as they were before the Great Fall. "They arrived by Fire and by Fire they will return," he ended by saying.

I was having a hard time with the odor of kerosene that exuded from the bonfire. Wondering how I would be able to last the night, I saw the dwarf put down his torch. He played a sweet melody on his drum which eased my focus on the smell. The group formed a circle and intoned chants and litanies directed toward the Fire Spirit. El Anciano, with his golden serpent head, accompanied by the rhythm of the drum, circled around the bonfire throwing into it powder from a black velvet pouch. As the sounds got louder, the nine began to dance, their masks balanced up on their heads.

The rhythm was so seductive that I caught myself swaying to it. Unable to stop, I started to accompany the chants in a low voice. The dwarf played louder and louder and, without stopping the dancing, everyone gathered closer to the fire. The candle flames, flickering on the trees, produced an agreeable sensation which blended with the animation of the group. The mood was contagious.

The dancers were turning counterclockwise while El Anciano, a little apart from the rest, went on throwing the strange powder into the fire and making mysterious signs in the air. Everyone sang still louder and the powerful rhythm increased. The Homage to the Crystals was carried out several times. I could see masked faces, flickering candles and the sign of the diamond over their heads.

With the chants reaching a great crescendo, El

Anciano abruptly grabbed his torch from the ground and threw it, soaked with kerosene, into the fire. Everyone jumped back as a huge flame shot up several yards, roaring into the night. The leaves in the clearing rustled, stirred up by the force of the air from the blaze. With arms raised, everyone acclaimed that the fire had succeeded in conquering the dark, bathing the meadow and the clouds above in a crimson glow.

The drum did not stop and everyone came forward to toss their candles into the great fire. Sparks rained down all around. The hills had assumed a red hue, the reflection of flames dancing upon them in soft waves. While El Anciano shouted sacred names and invoked the Spirit, the group continued whirling around with chants becoming shouts, sonorous shouts that called for the presence of the Fire Spirit. The shouts competed with stomping feet that slapped the ground. The dwarf was clapping and jumping even more frenetically, perhaps compensating for the drum now losing its beat.

Then, El Anciano opened a box and took out something that he tossed into the fire. This was a huge crystal geode. It landed like a shot from a gun and made the flames crackle violently.

An immense flame arose above the group, diminished, then returned with even more intensity. Within it I could see a face, a vision of an animal with cone-shaped eyes and well-defined features. It had an expression of defiance and domination. I knew this was something unreal. The Brothers of Fire stood still with arms raised and sent out a long, fierce moan toward the majestic image that hovered in the air.

Next, a kind of poltergeist phenomenon sent my crystal flying from my hands. My hair stood on end, and my body became intolerably hot from head to toe. I felt like a five-pointed star, with my extremities burning and my eyes blank. But the heat left as quickly as it had come, leaving me with the sensation of being very light.

Suddenly, as if we had traveled to another dimension, all noise ceased. The atmosphere was filled with calm and serenity. Everything began to turn slowly in circles—the people, the flames and even the air. As if in slow motion, I saw everyone remove their tunics and masks and hand them over to the extraordinary energy emitted by the Fire Spirit. I gazed at the stars and infinite sky, enjoying a profound sensation of peace.

As if in a dream, I watched the final stage of the ceremony. The Brothers went by the path back to the open clearing, talking animatedly and united by a great love. They had left a pig roasting on the fire which they now ate with wine served by the dwarf. They feasted and drank throughout the night and then everyone departed at dawn.

Feeling marvelously well, I returned home and slept all day. In the evening, I went searching for El Anciano. When I found him, we exchanged not one word; there wasn't anything to say. Just by looking at each other, we understood everything. Now that I had seen the Fire Spirit, there was nothing else to learn. Everything was clear, incredibly clear. The figure that had appeared above the enormous flame spoke for itself, and it would forever remain in my eyes, my soul and my heart.

Afterword

A few days later, the foreman came in a car to return me to Rio de Janeiro. El Anciano and I said good-bye with only a silent look. The teacher was content with his disciple, the disciple thankful for his teacher. We embraced affectionately with the sadness of parting alleviated by the hope of meeting again soon.

In spite of our separation, I knew that I carried El Anciano in my heart and, spiritually, we would always remain together. I left his place completely different from when I had arrived. A whole new world had been revealed to me. I was certain I would never forget the expression of the Fire Spirit, a presence that had strengthened me and helped me to learn, assimilate and understand what was important in life.

The program works!

ABOUT THE AUTHOR

The young Brazilian writer, Francisco Boström, is one of the most important personalities of the new consciousness movement of Brazil. From a traditional family of Swedish descent, Boström is an expert in precious stones and ran his own company—Brazilian Crystals Ltda—until he encountered *El Anciano*, "The Wise Old One." His life took a dramatic turn when he became a member of the secret Brotherhood of Fire. This esoteric society teaches magical alchemy using natural crystals and fire. Francisco Boström has now dedicated himself to spreading the wisdom he has learned from *El Anciano* and travels internationally, giving lectures and seminars.

BOOKS FROM MERRILL-WEST PUBLISHING

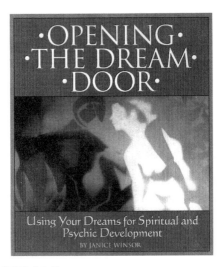

·OPENING·
·THE DREAM·
·DOOR·

Using Your Dreams for Spiritual and
Psychic Development
BY JANICE WINSOR

ISBN 1-886708-04-5
160 pages, $14.95
Janice Winsor

Written in a friendly personal style, *Opening the Dream Door* does not attempt to interpret dream symbology. The author believes that each person has their own personal mythology and symbology. By using her own dreams as an example, Janice Winsor makes practical suggestions on expanding your subconscious and connecting to other realms. There are chapters that address different types of supernatural connections, such as how to channel, heal your body, connect with animal allies, spirit guides, past lives and alien encounters. This is a fun, enjoyable manual for awakening psychic and intuitive gifts through the dream world.

Janice Windsor is a certified Clinical Hypnotherapist and spiritual counselor. A native of California, she is also an artist and writer of numerous books. She is presently residing in Australia—her new home of the heart.

Books from Merrill-West Publishing

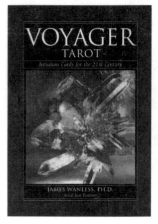

Voyager is a timeless symbology beyond any tarot deck. Experience the power of symbols...Pick a card a day and change your life!

Voyager Tarot® shows you how to achieve your visions and dreams, creating the succes you desire. Use the deck as an intuitive guide to forecast your future, reveal secrets of your subconscious and provide spiritual inspiration. This excellent decision making tool can help navigate life, enhancing business planning, relationship building and self-discovery.

ISBN 1-886708-05-3 Box
Book and card set: $35.00
James Wanless, Ph.D. & Artist, Ken Knutson

James Wanless, Ph.D., also known as "Mr. Tarot," received his doctorate in political science from Columbia University in 1972. His research led him to a path of "inner peace tradition." Traveling the far corners of the earth, he has pursued his study of symbols and myths, shamanism, and vision-making. He is the author of *Strategic Intuition for the 21st Century: Tarot for Business* and numerous books on intuition and tarot.

BOOKS FROM MERRILL-WEST PUBLISHING

Voyager Tarot Way of the Great Oracle, companion book to the highly acclaimed Voyager Tarot Deck, brings together symbols and signs so that anyone can become their own oracle and realize their highest dreams. Dr. Wanless also shows how the Runes, Auras, Crystals, Palms, Shamanism, etc. fit into our universe.

ISBN 1-886708-06-1
330 pages, $14.95
James Wanless, Ph.D.

This combination workbook, instruction manual and glossary of symbols shows you how to use tarot to facilitate intuition, healing, decision-making, play, relationships, prosperity, meditation, self-affirmation and spiritual growth. Includes complete interpretations of the 78 cards of the Thoth deck. It also contains 28 original layouts for reading the cards plus 22 different ways of using the cards.

ISBN 0-9615079-1-8
164 pages, $12.95
James Wanless, Ph.D.

For a retail or wholesale catalog

Merrill - West Publishing
PO Box 1227, Carmel, Ca. 93921, USA
1-800-676-1256 or 831-644-9096
fax: 831-644-9097
e-mail info@voyagertarot.com
www.voyagertarot.com

(area-code is 408 until July 1, 1998 then changes to 831)